RISK MANAGEMENT PROGRAM GUIDANCE FOR OFFSITE CONSEQUENCE ANALYSIS

This document provides guidance to the owner or operator of processes covered by the Chemical Accident Prevention Program rule in the analysis of offsite consequences of accidental releases of substances regulated under section 112(r) of the Clean Air Act. This document does not substitute for EPA's regulations, nor is it a regulation itself. Thus, it cannot impose legally binding requirements on EPA, States, or the regulated community, and may not apply to a particular situation based upon the circumstances. This guidance does not constitute final agency action, and EPA may change it in the future, as appropriate.

TABLE OF CONTENTS

TABLE OF CONTENTS
(Continued)

Chapter | Page

TABLE OF CONTENTS
(Continued)

TABLE OF CONTENTS
(Continued)

TABLE OF CONTENTS
(Continued)

APPENDICES Page

TABLE OF CONTENTS
(Continued)

APPENDICES | Page

LIST OF EXHIBITS

TABLE OF POTENTIALLY REGULATED ENTITIES

This table is not intended to be exhaustive, but rather provides a guide for readers regarding entities likely to be regulated under 40 CFR part 68. This table lists the types of entities that EPA is now aware could potentially be regulated by this rule (see Appendix B of the "General Guidance for Risk Management Programs" for a more detailed list of potentially affected NAICS codes). Other types of entities not listed in this table could also be affected. To determine whether your facility is covered by the risk management program rules in part 68, you should carefully examine the applicability criteria discussed in Chapter 1 of the General Guidance and in 40 CFR 68.10. If you have questions regarding the applicability of this rule to a particular entity, call the EPCRA/CAA Hotline at (800) 424-9346 (TDD: (800) 553-7672).

Category	NAICS Codes	SIC Codes	Examples of Potentially Regulated Entities
Chemical manufacturers	325	28	Petrochemicals Industrial gas Alkalies and chlorine Industrial inorganics Industrial organics Plastics and resins Agricultural chemicals Soap, cleaning compounds Explosives Miscellaneous chemical manufacturing
Petroleum refineries	32411	2911	Petroleum refineries
Pulp and paper	322	26	Paper mills Pulp mills Paper products
Food processors	311	20	Dairy products Fruits and vegetables Meat products Seafood products
Polyurethane foam	32615	3086	Plastic foam products
Non-metallic mineral products	327	32	Glass and glass products Other non-metallic mineral products
Metal products	331 332	33 34	Primary metal manufacturing Fabricated metal products

Category	NAICS Codes	SIC Codes	Examples of Potentially Regulated Entities
Machinery manufacturing	333	35	Industrial machinery Farm machinery Other machinery
Computer and electronic equipment	334	36	Electronic equipment Semiconductors
Electric equipment	335	36	Lighting Appliance manufacturing Battery manufacturing
Transportation equipment	336	37	Motor vehicles and parts Aircraft
Food distributors	4224 4228	514 518	Frozen and refrigerated foods Beer and wines
Chemical distributors	42269	5169	Chemical wholesalers
Farm supplies	42291	5191	Agricultural retailers and wholesalers
Propane dealers	454312	5171 5984	Propane retailers and wholesalers
Warehouses	4931	422	Refrigerated warehouses Warehouse storing chemicals
Water treatment	22131	4941	Drinking water treatment systems
Wastewater treatment	22132 56221	4952 4933	Sewerage systems Wastewater treatment Waste treatment
Electric utilities	22111	4911	Electric power generation
Propane users			Manufacturing facilities Large institutions Commercial facilities
Federal facilities			Military installations Department of Energy installations

Roadmap to Offsite Consequence Analysis Guidance by Type of Chemical

Type of Chemical and Release Scenario	Applicable Sections and Appendices
Toxic Gas	
Worst-Case Scenario	
1) Define Worst Case	Section 2.1
2) Select Scenario	Sections 2.2 and 2.3
3) Calculate Release Rates	
Unmitigated	Section 3.1.1
Passive Mitigation	Section 3.1.2
Refrigerated	Section 3.1.3
4) Find Toxic Endpoint	Appendix B (Exhibit B-1)
5) Determine Reference Table and Distance	Section 3.1.3, 3.2.3
Dense or Neutrally Buoyant Plume	Chapter 4 and Appendix B (Exhibit B-1)
Chemical-Specific Tables (ammonia, chlorine, sulfur dioxide)	Chapter 4
Urban or Rural	Section 2.1 and Chapter 4
Release Duration	Section 2.1
Alternative Scenario	
1) Define Alternative Scenario	Chapter 6
2) Select Scenario	Chapter 6
3) Calculate Release Rates	
Unmitigated (from tanks and pipes)	Section 7.1.1
Active or Passive Mitigation	Section 7.1.2
4) Find Toxic Endpoint	Appendix B (Exhibit B-1)
5) Determine Reference Table and Distance	
Dense or Neutrally Buoyant Plume	Chapter 8 and Appendix B (Exhibit B-1)
Chemical-Specific Tables (ammonia, chlorine, sulfur dioxide)	Chapter 8
Urban or Rural	Section 2.1 and Chapter 8
Release Duration	Section 7.1

Roadmap to Offsite Consequence Analysis Guidance by Type of Chemical (continued)

Type of Chemical and Release Scenario	Applicable Sections and Appendices
Toxic Liquid	
Worst-Case Scenario	
1) Define Worst Case	Section 2.1
2) Select Scenario	Sections 2.2 and 2.3
3) Calculate Release Rates	
Releases from Pipes	Section 3.2.1
Unmitigated Pool Evaporation	Section 3.2.2
Passive Mitigation (dikes, buildings)	Section 3.2.3
Release at Ambient Temperature	Section 3.2.2, 3.2.3
Release at Elevated Temperature	Section 3.2.2, 3.2.3
Releases of Mixtures	Section 3.2.4 and Appendix B (Section B.2)
Temperature Corrections for Liquids at 25-50 °C	Section 3.2.5 and Appendix B (Exhibit B-4)
Releases of Solutions	Section 3.3 and Appendix B (Exhibit B-3)
4) Find Toxic Endpoint	
For Liquids/Mixtures	Appendix B (Exhibit B-2)
For Solutions	Appendix B (Exhibit B-3)
5) Determine Reference Table and Distance	
Dense or Neutrally Buoyant Plume (liquids)	Chapter 4 and Appendix B (Exhibit B-2)
Dense or Neutrally Buoyant Plume (solutions)	Chapter 4 and Appendix B (Exhibit B-3)
Chemical Specific Table (aqueous ammonia)	Chapter 4
Urban or Rural	Section 2.1 and Chapter 4
Release Duration (liquids)	Section 3.2.2
Release Duration (solutions)	Chapter 4

Roadmap to Offsite Consequence Analysis Guidance by Type of Chemical (continued)

Type of Chemical and Release Scenario	Applicable Sections and Appendices
Toxic Liquid	
Alternative Scenario	
1) Define Alternative Scenario	Chapter 6
2) Select Scenario	Chapter 6
3) Calculate Release Rates	Section 7.2
Unmitigated (from tanks and pipes)	Section 7.2.1
Active or Passive Mitigation	Section 7.2.2
Release at Ambient Temperature	Section 7.2.3
Release at Elevated Temperature	Section 7.2.3
Release of Solution	Sections 7.2.4 and 3.3 and Appendix B (Exhibit B-3)
4) Find Toxic Endpoint	
For Liquids/Mixtures	Appendix B (Exhibit B-2)
For Solutions	Appendix B (Exhibit B-3)
5) Determine Reference Table and Distance	
Dense or Neutrally Buoyant Plume (liquids/mixtures)	Chapter 8 and Appendix B (Exhibit B-2)
Dense or Neutrally Buoyant Plume (solutions)	Chapter 8 and Appendix B (Exhibit B-3)
Chemical-Specific Table (aqueous ammonia)	Chapter 8
Urban or Rural	Section 2.1 and Chapter 8
Release Duration (liquids/mixtures)	Section 7.2
Release Duration (solutions)	Chapter 8

Roadmap to Offsite Consequence Analysis Guidance by Type of Chemical (continued)

Type of Chemical and Release Scenario	Applicable Sections and Appendices
Flammable Substance	
Worst-Case Scenario	
1) Define Worst Case	Sections 5.1 and 2.1
2) Select Scenario	Sections 5.1, 2.2, and 2.3
3) Determine Distance to Overpressure Endpoint	
For Pure Flammable Substances	Section 5.1
For Flammable Mixtures	Section 5.2
Alternative Scenario	
1) Define Alternative Scenario	Chapter 6
2) Select Scenario	Chapter 6
3) For Vapor Cloud Fires	
Calculate Release Rates (gases)	Section 9.1 and Appendix C (Exhibit C-2)
Calculate Release Rates (liquids)	Section 9.2 and Appendix C (Exhibit C-3)
Find Lower Flammability Limit (gases)	Appendix C (Exhibit C-2)
Find Lower Flammability Limit (liquids)	Appendix C (Exhibit C-3)
Dense or Neutrally Buoyant (gases)	Appendix C (Exhibit C-2)
Dense or Neutrally Buoyant (liquids)	Appendix C (Exhibit C-3)
Urban or Rural	Section 10.1
Release Duration	Section 10.1
Determine Distance	Section 10.1
4) For Pool Fires	Section 10.2 and Appendix C (Exhibit C-3)
5) For BLEVEs	Section 10.3
6) For Vapor Cloud Explosions	Section 10.4

This page intentionally left blank.

1 INTRODUCTION

1.1 Purpose of this Guidance

This document provides guidance on how to conduct the offsite consequence analyses for Risk Management Programs required under the Clean Air Act (CAA). Section 112(r)(7) of the CAA directed the U.S. Environmental Protection Agency (EPA) to issue regulations requiring facilities with large quantities of very hazardous chemicals to prepare and implement programs to prevent the accidental release of those chemicals and to mitigate the consequences of any releases that do occur. EPA issued that rule, "Chemical Accident Prevention Provisions" on June 20, 1996. The rule is codified at part 68 of Title 40 of the Code of Federal Regulations (CFR). If you handle, manufacture, use, or store any of the toxic or flammable substances listed in 40 CFR 68.130 above the specified threshold quantities in a process, you are required to develop and implement a risk management program under part 68 of 40 CFR. The rule applies to a wide variety of facilities that handle, manufacture, store, or use toxic substances, including chlorine and ammonia, and highly flammable substances, such as propane. If you are not sure whether you are subject to the rule, you should review the rule and Chapters 1 and 2 of EPA's *General Guidance for Risk Management Programs (40 CFR part 68)*, available from EPA at www.epa.gov/emergencies/rmp.

If you are subject to the rule, you are required to conduct an offsite consequence analysis to provide information to the state, local, and federal governments and the public about the potential consequences of an accidental chemical release. The offsite consequence analysis consists of two elements:

- A worst-case release scenario, and
- Alternative release scenarios.

To simplify the analysis and ensure comparability, EPA has defined the worst-case scenario as the release of the largest quantity of a regulated substance from a single vessel or process line failure that results in the greatest distance to an endpoint. In broad terms, the distance to the endpoint is the distance a toxic vapor cloud, heat from a fire, or blast waves from an explosion will travel before dissipating to the point that serious injuries from short-term exposures will no longer occur. Endpoints for regulated substances are specified in 40 CFR 68.22(a) and Appendix A of part 68 and are presented in Appendices B and C of this guidance.

Alternative release scenarios are scenarios that are more likely to occur than the worst-case scenario and that will reach an endpoint offsite, unless no such scenario exists. Within these two parameters, you have flexibility to choose alternative release scenarios that are appropriate for your site. The rule, in 40 CFR 68.28 (b)(2), and the *General Guidance for Risk Management Programs (40 CFR part 68)*, Chapter 4, provide examples of alternative release scenarios that you should consider when conducting the offsite consequence analysis.

RMP*Comp™

To assist those using this guidance, the National Oceanic and Atmospheric Administration (NOAA) and EPA have developed a software program, RMP*Comp™, that performs the calculations described in this document. This software can be downloaded from the EPA/OEM website at www.epa.gov/emergencies/rmp.

This guidance document provides a simple methodology for conducting offsite consequence analyses. You may use simple equations to estimate release rates and reference tables to determine distances to the endpoint of concern. This guidance provides generic reference tables of distances, applicable to most of the regulated toxic substances, and chemical-specific tables for ammonia, chlorine, and sulfur dioxide. This guidance also provides reference tables of distances for consequences of fires and explosions of flammable substances. In some cases, the rule allows users of this document to adopt generic assumptions rather than the site-specific data required if another model is employed (see Exhibit 1).

The methodology and reference tables of distances presented here are optional. You are not required to use this guidance. You may use publicly available or proprietary air dispersion models to do your offsite consequence analysis, subject to certain conditions. If you choose to use models instead of this guidance, you should review the rule and Chapter 4 of the General Guidance for Risk Management Programs, which outline required conditions for use of models. In selected example analyses, this document presents the results of some models to provide a basis for comparison. It also indicates certain conditions of a release that may warrant more sophisticated modeling than is represented here. However, this guidance does not discuss the procedures to follow when using models; if you choose to use models, you should consult the appropriate references or instructions for those models.

This guidance provides distances to endpoints for toxic substances that range from 0.1 miles to 25 miles. Other models may not project distances this far (and some may project even longer distances). One commonly used model, ALOHA, has an artificial distance cutoff of 6 miles (i.e., any scenario which would result in an endpoint distance beyond 6 miles is reported as "greater than 6 miles"). Although you may use ALOHA if it is appropriate for the substance and scenario, you should consider choosing a different model if the scenario would normally result in an endpoint distance significantly greater than 6 miles. Otherwise, you should be prepared to explain the difference between your results and those in this guidance or other commonly used models. Also, you should be aware that the RMP*eSubmit system accepts only numerical entries (i.e., it will not accept a "greater than" distance). If you do enter a distance in RMP*eSubmit that is the result of a particular model's maximum distance cutoff (including the maximum distance cutoff in this guidance), you can explain this in the executive summary of your RMP.

Exhibit 1
Required Parameters for Modeling (40 CFR 68.22)

WORST CASE	ALTERNATIVE SCENARIO
Endpoints (§68.22(a))	
Endpoints for toxic substances are specified in part 68 Appendix A.	Endpoints for toxic substances are specified in part 68 Appendix A.
For flammable substances, endpoint is overpressure of 1 pound per square inch (psi) for vapor cloud explosions.	For flammable substances, endpoint is: ♦ Overpressure of 1 psi for vapor cloud explosions, or ♦ Radiant heat level of 5 kilowatts per square meter (kW/m^2) for 40 seconds for heat from fires (or equivalent dose), or ♦ Lower flammability limit (LFL) as specified in NFPA documents or other generally recognized sources.
Wind speed/stability (§68.22(b))	
This guidance assumes 1.5 meters per second and F stability. For other models, use wind speed of 1.5 meters per second and F stability class unless you can demonstrate that local meteorological data applicable to the site show a higher minimum wind speed or less stable atmosphere at all times during the previous three years. If you can so demonstrate, these minimums may be used for site-specific modeling.	This guidance assumes wind speed of 3 meters per second and D stability. For other models, you must use typical meteorological conditions for your site.
Ambient temperature/humidity (§68.22(c))	
This guidance assumes 25°C (77°F) and 50 percent humidity. For other models for toxic substances, you must use the highest daily maximum temperature and average humidity for the site during the past three years.	This guidance assumes 25°C and 50 percent humidity. For other models, you may use average temperature/humidity data gathered at the site or at a local meteorological station.
Height of release (§68.22(d))	
For toxic substances, you must assume a ground level release.	This guidance assumes a ground-level release. For other models, release height may be determined by the release scenario.
Surface roughness (§68.22(e))	
Use urban (obstructed terrain) or rural (flat terrain) topography, as appropriate.	Use urban (obstructed terrain) or rural (flat terrain) topography, as appropriate.
Dense or neutrally buoyant gases (§68.22(f))	
Tables or models used for dispersion of regulated toxic substances must appropriately account for gas density. If you use this guidance, see Tables 1-4 for neutrally buoyant gases and Tables 5-8 for dense gases, or Tables 9-12 for specific chemicals.	Tables or models used for dispersion must appropriately account for gas density. If you use this guidance, see Tables 14-17 for neutrally buoyant gases and Tables 18-21 for dense gases, or Tables 22-25 for specific chemicals.
Temperature of released substance (§68.22(g))	
You must consider liquids (other than gases liquefied by refrigeration) to be released at the highest daily maximum temperature, from data for the previous three years, or at process temperature, whichever is higher. Assume gases liquefied by refrigeration at atmospheric pressure to be released at their boiling points. This guidance provides factors for estimation of release rates at 25°C or the boiling point of the released substance, and also provides temperature correction factors.	Substances may be considered to be released at a process or ambient temperature that is appropriate for the scenario. This guidance provides factors for estimation of release rates at 25°C or the boiling point of the released substance, and also provides temperature correction factors.

1.2 This Guidance Compared to Other Models

Results obtained using the methods in this document are expected to be conservative (i.e., they will generally, but not always, overestimate the distance to endpoints). The chemical-specific reference tables in this guidance provide less conservative results than the generic reference tables, because the chemical-specific tables were derived using more realistic assumptions and considering more factors.

Complex models that can account for many site-specific factors may give less conservative estimates of offsite consequences than the simple methods in this guidance. This is particularly true for alternative scenarios, for which EPA has not specified many assumptions. However, complex models may be expensive and require considerable expertise to use; this guidance is designed to be simple and straightforward. You will need to consider these tradeoffs in deciding how to carry out your required consequence analyses. Appendix A provides information on references for some other methods of analysis; these references do not include all models that you may use for these analyses. You will find that modeling results will sometimes vary considerably from model to model.

1.3 Number of Scenarios to Analyze

The number and type of analyses you must perform depend on the "Program" level of each of your processes. The rule defines three Program levels. Processes are eligible for Program 1 if, among other criteria, there are no public receptors within the distance to the endpoint for the worst-case scenario. Because no public receptors would be affected by the worst-case release, no further modeling is required for these processes. For processes subject to Program 2 or Program 3, both worst-case release scenarios and alternative release scenarios are required. To determine the Program level of your processes, consult 40 CFR 68.10(b), (c), and (d), or Chapter 2 of EPA's *General Guidance for Risk Management Programs (40 CFR part 68)*.

Once you have determined the Program level of your processes, you are required to conduct the following offsite consequence analyses:

- One worst-case release scenario for each Program 1 process;
- One worst-case release scenario to represent all regulated toxic substances in Program 2 and Program 3 processes;
- One worst-case release scenario to represent all regulated flammable substances in Program 2 and Program 3 processes;
- One alternative release scenario for each regulated toxic substance in Program 2 and Program 3 processes; and
- One alternative release scenario to represent all regulated flammable substances in Program 2 and Program 3 processes.

NOTE: You may need to analyze additional worst-case scenarios if release scenarios for regulated flammable or toxic substances from other covered processes at your facility would affect different public receptors. For example, worst-case release scenarios for storage tanks at opposite ends of your facility may potentially reach different areas where people could be affected. In that case, you will have to conduct analyses of and report on both releases.

GUIDANCE FOR INDUSTRY-SPECIFIC RISK MANAGEMENT PROGRAMS

EPA developed guidance for industry-specific risk management programs for the following industries:

- Propane storage facilities
- Chemical distributors
- Waste water treatment plants
- Warehouses
- Ammonia refrigeration
- Small propane retailers & users

Industry-specific guidance is either appended to this guidance or is contained in stand-alone documents that you can obtain from EPA's website at www.epa.gov/emergencies/rmp. If an industry-specific appendix or guidance document exists for your process(es), you should consider using it because it will provide more information that is specific to your process(es), including dispersion modeling and prevention program elements.

1.4 Modeling Issues

The consequences of an accidental chemical release depend on the conditions of the release and the conditions at the site at the time of the release. This guidance provides reference tables of distances, based on results of modeling, for estimation of worst-case and alternative scenario consequence distances. Worst-case consequence distances obtained using these tables are not intended to be precise predictions of the exact distances that might be reached in the event of an actual accidental release. For this guidance, worst-case distances are based on modeling results assuming the combination of worst-case conditions required by the rule. This combination of conditions occurs rarely and is unlikely to persist for very long. To derive the alternative scenario distances, less conservative assumptions were used for modeling; these assumptions were chosen to represent more likely conditions than the worst-case assumptions. Nevertheless, in an actual accidental release, the conditions may be very different. Users of this guidance should remember that the results derived from the methods presented here are rough estimates of potential consequence distances. Other models may give different results; the same model also may give different results if different assumptions about release conditions and/or site conditions are used.

The reference tables of distances in this guidance provide results to a maximum distance of 25 miles. EPA recognizes that modeling results at such large distances are highly uncertain. Almost no experimental data or data from accidents are available at such large distances to compare to modeling results. Most data are reported for distances well under 10 miles. Modeling uncertainties are likely to increase as distances increase because conditions (e.g., atmospheric stability, wind speed, surface roughness) are not likely to remain constant over large distances. Thus, at large distances (e.g., greater than about 6 to 10 miles), the modeling results should be viewed as very coarse estimates of consequence distances. EPA believes, however, that the results, even at large distances, can provide useful information for comparison purposes. For example, Local Emergency Planning Committees (LEPCs) and other local agencies can use relative differences in distance to aid in establishing chemical accident prevention and preparedness priorities among facilities in a community. Since worst-case scenario distances are based on modeling conditions that are unlikely to occur, and since modeling of any scenario that results in large distances is very uncertain, EPA strongly urges communities and industry not to rely on the results of worst-case modeling or any modeling that results in very large toxic endpoint distances in emergency planning and response activities. Results of alternative scenario models are apt to provide a more reasonable basis for planning and response.

1.5 Steps for Performing the Analysis

This Chapter presents the steps you should follow in using this guidance to carry out an offsite consequence analysis. Before carrying out one or more worst-case and/or alternative release analyses, you will need to obtain several pieces of information about the regulated substances you have, the area surrounding your site, and typical meteorological conditions:

- Determine whether each regulated substance is toxic or flammable, as indicated in the rule or Appendices B and C of this guidance.

- For the worst-case analysis, determine the quantity of each substance held in the largest single vessel or pipe.

- Collect information about any passive or active (alternative scenarios only) release mitigation measures that are in place for each substance.

- For toxic substances, determine whether the substance is stored as a gas, as a liquid, as a gas liquefied by refrigeration, or as a gas liquefied under pressure. For alternative scenarios involving a vapor cloud fire, you may also need this information for flammable substances.

- For toxic liquids, determine the highest daily maximum temperature of the liquid, based on data for the previous three years, or process temperature, whichever is higher.

- For toxic substances, determine whether the substance behaves as a dense or neutrally buoyant gas or vapor (see Appendix B, Exhibits B-1 and B-2). For alternative scenarios involving a vapor cloud fire, you will also need this information for flammable substances (see Appendix C, Exhibits C-2 and C-3).

- For toxic substances, determine whether the topography (surface roughness) of your site is either urban or rural as thse terms are defined by the rule (see 40 CFR 68.22(e)). For alternative scenarios involving a vapor cloud fire, you will also need this information for flammable substances.

After you have gathered the above information, you will need to take three steps (except for flammable worst-case releases):

(1) Select a scenario;

(2) Determine the release or volatilization rate; and
(3) Determine the distance to the endpoint.

For flammable worst-case scenarios, only steps one and three are needed. Sections 1.5.1 through 1.5.6 outline the procedures to perform the analyses. In addition to basic procedures, these sections provide references to sections of this guidance where you will find detailed instructions on carrying out the applicable portion of the analysis. Sections 1.5.1 through 1.5.3 below provide basic steps to analyze worst-case scenarios for toxic gases, toxic liquids, and flammable substances. Sections 1.5.4 through 1.5.6 provide basic steps for alternative scenario analysis. Appendix E of this document provides worksheets that may help you to perform the analyses.

1.5.1 Worst-Case Analysis for Toxic Gases

To conduct worst-case analyses for toxic gases, including toxic gases liquefied by pressurization (see Appendix E, Worksheet 1, for a worksheet that can be used in carrying out this analysis):

Step 1: Determine worst-case scenario. Identify the toxic gas, quantity, and worst-case release scenario, as defined by the rule (Chapter 2).

Step 2: Determine release rate. Estimate the release rate for the toxic gas, using the parameters required by the rule. This guidance provides methods for estimating the release rate for:

- Unmitigated releases (Section 3.1.1).
- Releases with passive mitigation (Section 3.1.2).

Step 3: Determine distance to endpoint. Estimate the worst-case consequence distance based on the release rate and toxic endpoint (defined by the rule) (Chapter 4). This guidance provides reference tables of distances (Reference Tables 1-12). Select the appropriate reference table based on the density of the released substance, the topography of your site, and the duration of the release (always 10 minutes for gas releases). Estimate distance to the endpoint from the appropriate table.

1.5.2 Worst-Case Analysis for Toxic Liquids

To conduct worst-case analyses for toxic substances that are liquids at ambient conditions or for toxic gases that are liquefied by refrigeration alone (see Appendix E, Worksheet 2, for a worksheet for this analysis):

Step 1: Determine worst-case scenario. Identify the toxic liquid, quantity, and worst-case release scenario, as defined by the rule (Chapter 2). To estimate the quantity of liquid released from piping, see Section 3.2.1.

Step 2: Determine release rate. Estimate the volatilization rate for the toxic liquid and the duration of the release, using the parameters required by the rule. This guidance provides methods for estimating the pool evaporation rate for:

- Gases liquefied by refrigeration alone (Sections 3.1.3 and 3.2.3).

- Unmitigated releases (Section 3.2.2).

- Releases with passive mitigation (Section 3.2.3).

- Releases at ambient or elevated temperature (Sections 3.2.2, 3.2.3, and 3.2.5).

- Releases of mixtures of toxic liquids (Section 3.2.4).

- Releases of common water solutions of regulated substances and of oleum (Section 3.3).

Step 3: Determine distance to endpoint. Estimate the worst-case consequence distance based on the release rate and toxic endpoint (defined by the rule) (Chapter 4). This guidance provides reference tables of distances (Reference Tables 1-12). Select the appropriate reference table based on the density of the released substance, the topography of your site, and the duration of the release. Estimate distance to the endpoint from the appropriate table.

1.5.3 Worst-Case Analysis for Flammable Substances

To conduct worst-case analyses for all regulated flammable substances (i.e., gases and liquids) (see Appendix E, Worksheet 3, for a worksheet for this analysis):

Step 1: Determine worst-case scenario. Identify the appropriate flammable substance, quantity, and worst-case scenario, as defined by the rule (Chapter 2).

Step 2: Determine distance to endpoint. Estimate the distance to the required overpressure endpoint of 1 psi for a vapor cloud explosion of the flammable substance, using the assumptions required by the rule (Chapter 5). This guidance provides a reference table of distances (Reference Table 13) for worst-case vapor cloud explosions. Estimate the distance to the endpoint from the quantity released and the table.

1.5.4 Alternative Scenario Analysis for Toxic Gases

To conduct alternative release scenario analyses for toxic gases, including toxic gases liquefied by pressurization (see Appendix E, Worksheet 4, for a worksheet for this analysis):

Step 1: Select alternative scenario. Choose an appropriate alternative release scenario for the toxic gas. This scenario should have the potential for offsite impacts unless no such scenario exists. (Chapter 6).

Step 2: Determine release rate. Estimate the release rate and duration of the release of the toxic gas, based on your scenario and site-specific conditions. This guidance provides methods for:

- Unmitigated releases (Section 7.1.1).

- Releases with active or passive mitigation (Section 7.1.2).

Step 3: Determine distance to endpoint. Estimate the alternative scenario distance based on the release rate and toxic endpoint (Chapter 8). This guidance provides reference tables of distances (Reference Tables 14-25) for alternative scenarios for toxic substances. Select the appropriate reference table based on the density of the released substance, the topography of your site, and the duration of the release. Estimate distance to the endpoint from the appropriate table.

1.5.5 Alternative Scenario Analysis for Toxic Liquids

To conduct alternative release scenario analyses for toxic substances that are liquids at ambient conditions or for toxic gases that are liquefied by refrigeration alone (see Appendix E, Worksheet 5, for a worksheet for this analysis):

Step 1: Select alternative scenario. Choose an appropriate alternative release scenario and release quantity for the toxic liquid. This scenario should have the potential for offsite impacts (Chapter 6), unless no such scenario exists.

Step 2: Determine release rate. Estimate the release rate and duration of the release of the toxic liquid, based on your scenario and site-specific conditions. This guidance provides methods to estimate the liquid release rate and quantity of liquid released for:

- Unmitigated liquid releases (Section 7.2.1).
- Mitigated liquid releases (Section 7.2.2).

The released liquid is assumed to form a pool. This guidance provides methods to estimate the pool evaporation rate and release duration for:

- Unmitigated releases (Section 7.2.3).
- Releases with passive or active mitigation (Section 7.2.3).
- Releases at ambient or elevated temperature (Sections 7.2.3).
- Releases of common water solutions of regulated substances and of oleum (Section 7.2.4).

Step 3: Determine distance to endpoint. Estimate the alternative scenario distance based on the release rate and toxic endpoint (Chapter 8). This guidance provides reference tables of distances (Reference Tables 14-25) for alternative scenarios for toxic substances. Select the appropriate reference table based on the density of the released substance, the topography of your site, and the duration of the release. Estimate distance to the endpoint from the appropriate table.

1.5.6 Alternative Scenario Analysis for Flammable Substances

To conduct alternative release scenario analyses for all regulated flammable substances (i.e., gases and liquids) (see Appendix E, Worksheet 6, for a worksheet for this analysis):

Step 1: Select alternative scenario. Identify the flammable substance, and choose the quantity and type of event for the alternative scenario consequence analysis (Chapter 6).

Step 2: Determine release rate. Estimate the release rate to air of the flammable gas or liquid, if the scenario involves a vapor cloud fire (Section 9.1 for flammable gases, Section 9.2 for flammable liquids).

Step 3: Determine distance to endpoint. Estimate the distance to the appropriate endpoint (defined by the rule). This guidance provides methods for:

- Vapor cloud fires (Section 10.1 and Reference Tables 26-29); select the appropriate reference table based on the density of the released substance and the topography of your site, and estimate distance to the endpoint from the appropriate table.

- Pool fires (Section 10.2); estimate distance from the equation and chemical-specific factors provided.

- BLEVEs (Section 10.3 and Reference Table 30); estimate distance from the quantity of flammable substance and the table.

- Vapor cloud explosions (Section 10.4 and Reference Table 13); estimate quantity in the cloud from the equation and chemical-specific factors provided, and estimate distance from the quantity, the table, and a factor provided for alternative scenarios.

1.6 Additional Sources of Information

EPA's risk management program requirements may be found at 40 CFR part 68. The relevant sections were published in the Federal Register on January 31, 1994 (59 FR 4478) and June 20, 1996 (61 FR 31667). Final rules amending the list of substances and thresholds were published on August 25, 1997 (62 FR 45130) and January 6, 1998 (63 FR 640). A consolidated copy of these regulations is available in Appendix F.

EPA is working with industry and local, state, and federal government agencies to assist sources in complying with these requirements. For more information, refer to the General Guidance for Risk Management Programs Appendix E (Technical Assistance). Appendices C and D of the General Guidance also provide points of contact for EPA and Occupational Safety and Health Administration (OSHA) at the state and federal levels for your questions. Your LEPC also can be a valuable resource.

Finally, if you have access to the Internet, EPA has made copies of the rules, fact sheets, and other related materials available from EPA's website at www.epa.gov/emergencies/rmp. Please check the site regularly, as additional materials are posted when they become available. If you do not have access to the Internet, you can call EPA's hotline at (800) 424-9346.

2 DETERMINING WORST-CASE SCENARIOS

In Chapter 2

2.1 Definition of Worst-Case Scenario

A worst-case release is defined as:

- The release of the largest quantity of a regulated substance from a vessel or process line failure, and
- The release that results in the greatest distance to the endpoint for the regulated toxic or flammable substance.

You may take administrative controls into account when determining the largest quantity. Administrative controls are written procedures that limit the quantity of a substance that can be stored or processed in a vessel or pipe at any one time or, alternatively, procedures that allow the vessel or pipe to occasionally store larger than usual quantities (e.g., during shutdown or turnaround). Endpoints for regulated substances are specified in the rule (40 CFR 68.22(a), and Appendix A to part 68 for toxic substances). For the worst-case analysis, you do not need to consider the possible causes of the worst-case release or the probability that such a release might occur; the release is simply assumed to take place. You must assume all releases take place at ground level for the worst-case analysis.

This guidance assumes meteorological conditions for the worst-case scenario of atmospheric stability class F (stable atmosphere) and wind speed 1.5 meters per second (3.4 miles per hour). Ambient air temperature for this guidance is 25 °C (77 °F). If you use this guidance, you may assume this ambient temperature for the worst case, even if the maximum temperature at your site in the last three years is higher.

The rule provides two choices for topography, urban and rural. EPA (40 CFR 68.22(e)) has defined urban as many obstacles in the immediate area, where obstacles include buildings or trees. Rural, by EPA's definition, means there are no buildings in the immediate area, and the terrain is generally flat and unobstructed. Thus, if your site is located in an area with few buildings or other obstructions (e.g., hills, trees), you should assume open (rural) conditions. If your site is in an area with many obstructions, even if it is in a remote location that would not usually be considered urban, you should assume urban conditions.

Toxic Gases

Toxic gases include all regulated toxic substances that are gases at ambient temperature (25 °C, 77 °F), with the exception of gases liquefied by refrigeration under atmospheric pressure and released into diked areas. For the worst-case consequence analysis, you must assume that a gaseous release of the total quantity occurs in 10 minutes. You may take passive mitigation measures (e.g., enclosure) into account in the analysis of the worst-case scenario.

Gases liquefied by refrigeration alone and released into diked areas may be modeled as liquids at their boiling points and assumed to be released from a pool by evaporation (40 CFR 68.25(c)(2)). Gases liquefied by refrigeration alone that would form a pool one centimeter or less in depth upon release must be modeled as gases. (Modeling indicates that pools one centimeter or less deep formed by gases liquefied by refrigeration would completely evaporate in 10 minutes or less, giving a release rate that is equal to or greater than the worst-case release rate for a gaseous release. In this case, therefore, it is appropriate to treat these substances as gases for the worst-case analysis.)

Endpoints for consequence analysis for regulated toxic substances are specified in the rule (40 CFR part 68, Appendix A). Exhibit B-1 of Appendix B lists the endpoint for each toxic gas. These endpoints are used for air dispersion modeling to estimate the consequence distance.

Toxic Liquids

For toxic liquids, you must assume that the total quantity in a vessel is spilled. This guidance assumes the spill takes place onto a flat, non-absorbing surface. For toxic liquids carried in pipelines, the quantity that might be released from the pipeline is assumed to form a pool. You may take passive mitigation systems (e.g., dikes) into account in consequence analysis. The total quantity spilled is assumed to spread instantaneously to a depth of one centimeter (0.033 foot or 0.39 inch) in an undiked area or to cover a diked area instantaneously. The temperature of the released liquid must be the highest daily maximum temperature occurring in the past three years or the temperature of the substance in the vessel, whichever is higher (40 CFR 68.25(d)(2)). The release rate to air is estimated as the rate of evaporation from the pool. If liquids at your site might be spilled onto a surface that could rapidly absorb the spilled liquid (e.g., porous soil), the methods presented in this guidance may greatly overestimate the consequences of a release. Consider using another method in such a case.

Exhibit B-2 of Appendix B presents the endpoint for air dispersion modeling for each regulated toxic liquid (the endpoints are specified in 40 CFR part 68, Appendix A).

Flammable Substances

For all regulated flammable substances, you must assume that the worst-case release results in a vapor cloud containing the total quantity of the substance that could be released from a vessel or pipeline. For the worst-case consequence analysis, you must assume the vapor cloud detonates. If you use a TNT-equivalent method for your analysis, you must assume a 10 percent yield factor.

The rule specifies the endpoint for the consequence analysis of a vapor cloud explosion of a regulated flammable substance as an overpressure of 1 pound per square inch (psi). This endpoint was chosen as the threshold for potential serious injuries to people as a result of property damage caused by an explosion (e.g., injuries from flying glass from shattered windows or falling debris from damaged houses). (See Appendix D, Section D.5 for additional information on this endpoint.)

Effect of Required Assumptions

The assumptions required for the worst-case analysis are intended to provide conservative worst-case consequence distances, rather than accurate predictions of the potential consequences of a release; that is, in most cases your results will overestimate the effects of a release. In certain cases, actual conditions could be even more severe than these worst-case assumptions (e.g., very high process temperature, high process pressure, or unusual weather conditions, such as temperature inversions); in such cases, your results might underestimate the effects. However, the required assumptions generally are expected to give conservative results.

2.2 Determination of Quantity for the Worst-Case Scenario

EPA has defined a worst-case release as the release of the largest quantity of a regulated substance from a vessel or process line failure that results in the greatest distance to a specified endpoint. For substances in vessels, you must assume release of the largest amount in a single vessel. For substances in pipes, you must assume release of the largest amount in a pipe. The largest quantity should be determined taking into account administrative controls rather than absolute capacity of the vessel or pipe. Administrative controls are written procedures that limit the quantity of a substance that can be stored or processed in a vessel or pipe at any one time, or, alternatively, occasionally allow a vessel or pipe to store larger than usual quantities (e.g., during turnaround).

2.3 Selecting Worst-Case Scenarios

Under part 68, a worst-case release scenario analysis must be completed for all covered processes, regardless of program level. The number of worst-case scenarios you must analyze depends on several factors. You need to consider only the hazard (toxicity or flammability) for which a substance is regulated (i.e., even if a regulated toxic substance is also flammable, you only need to consider toxicity in your analysis; even if a regulated flammable substance is also toxic, you only need to consider flammability).

For every Program 1 process, you must report the worst-case scenario with the greatest distance to an endpoint. If a Program 1 process has more than one regulated substance held above its threshold, you must determine which substance produces the greatest distance to its endpoint and report on that substance. If a Program 1 process has both regulated toxics and flammables above their thresholds, you still report only the one scenario that produces the greatest distance to the endpoint. The process is eligible for Program 1 if there are no public receptors within the distance to an endpoint of the worst-case scenario for the process and the other Program 1 criteria are met. For Program 2 or Program 3 processes, you must analyze and report on one worst-case analysis representing all toxic regulated substances present above the threshold quantity and one worst-case analysis representing all flammable regulated substances present above the threshold quantity. You may need to submit an additional worst-case analysis if a worst-case release from elsewhere at the source would potentially affect public receptors different from those affected by the initial worst-case scenario(s).

If you have more than one regulated substance in a class, the substance chosen for the consequence analysis for each hazard for Program 2 and 3 processes should be the substance that has the potential to cause the greatest offsite consequences. Choosing the toxic regulated substance that might lead to the greatest offsite consequences may require a screening analysis of the toxic regulated substances on site, because the potential consequences are dependent on a number of factors, including quantity, toxicity, and volatility.

Location (distance to the fenceline) and conditions of processing or storage (e.g., a high temperature process) also should be considered. In selecting the worst-case scenario, you may want to consider the following points:

- Toxic gases with low toxic endpoints are likely to give the greatest distances to the endpoint for a given release quantity; a toxic gas would be a likely choice for the worst-case analysis required for Program 2 and 3 processes (processes containing toxic gases are unlikely to be eligible for Program 1).

- Volatile, highly toxic liquids (i.e., liquids with high ambient vapor pressure and low toxic endpoints) also are likely to give large distances to the endpoint (processes containing this type of substance are unlikely to be eligible for Program 1).

- Toxic liquids with relatively low volatility (low vapor pressure) and low toxicity (large toxic endpoint) in ambient temperature processes may give fairly small distances to the endpoint; you probably would not choose such substances for the worst-case analysis for Program 2 or 3 if you have other regulated toxics, but you may want to consider carrying out a worst-case analysis to demonstrate potential Program 1 eligibility.

For flammable substances, you must consider the consequences of a vapor cloud explosion in the analysis. The severity of the consequences of a vapor cloud explosion depends on the quantity of the released substance in the vapor cloud, its heat of combustion, and other factors that are assumed to be the same for all flammable substances. In most cases, the analysis probably should be based on the regulated flammable substance present in the greatest quantity; however, a substance with a high heat of combustion may have a greater potential offsite impact than a larger quantity of a substance with a lower heat of combustion. In some cases, a regulated flammable substance that is close to the fenceline might have a greater potential offsite impact than a larger quantity farther from the fenceline.

You are likely to estimate smaller worst-case distances for flammable substances than for similar quantities of most toxic substances. Because the distance to the endpoint may be relatively small, you may find it worthwhile to carry out a worst-case analysis for each process containing flammable substances to demonstrate potential eligibility for Program 1, unless there are public receptors close to the process.

3 RELEASE RATES FOR TOXIC SUBSTANCES

In Chapter 3

- 3.1 Estimation of worst-case release rates for toxic gases.
- 3.2 Estimation of release rates for toxic liquids evaporating from pools.
- 3.3 Estimation of release rates for common water solutions of toxic substances and for oleum.

This chapter describes simple methods for estimating release rates for regulated toxic substances for the worst-case scenario. Simple release rate equations are provided, and factors to be used in these equations are provided (in Appendix B) for each regulated substance. The estimated release rates may be used to estimate dispersion distances to the toxic endpoint for regulated toxic gases and liquids, as discussed in Chapter 4.

3.1 Release Rates for Toxic Gases

In Section 3.1

- 3.1.1 Method to estimate worst-case release rates for unmitigated releases (releases directly to the air) of toxic gas.
- 3.1.2 Method to estimate worst-case release rates for toxic gas in enclosures (passive mitigation).
- 3.1.3 Method to estimate worst-case release rates for liquefied refrigerated toxic gases in diked areas (as toxic liquid - see Section 3.2.3), including consideration of the duration of the release.

Regulated substances that are gases at ambient temperature (25 °C, 77 °F) should be considered gases for consequence analysis, with the exception of gases liquefied by refrigeration at atmospheric pressure. Gases liquefied under pressure should be treated as gases. Gases liquefied by refrigeration alone and released into diked areas may be treated as liquids at their boiling points if they would form a pool upon release that is more than one centimeter (0.033 foot) in depth. Gases liquefied by refrigeration alone that would form a pool one centimeter (0.033 foot) or less in depth should be treated as gases. Modeling shows that the evaporation rate from such a pool would be equal to or greater than the rate for a toxic gas, which is assumed to be released over 10 minutes; therefore, treating liquefied refrigerated gases as gases rather than liquids in such cases is reasonable. You may consider passive mitigation for gaseous releases and releases of gases liquefied by refrigeration.

3.1.1 Unmitigated Releases of Toxic Gas

If no passive mitigation system is in place, estimate the release rate for the release over a 10-minute period of the largest quantity resulting from a pipe or vessel failure, as required by the rule (40 CFR 68.25(c)). For a release from a vessel, calculate the release rate as follows:

$$QR = \frac{QS}{10} \tag{3-1}$$

where: QR = Release rate (pounds per minute)
QS = Quantity released (pounds)

Example 1. Gas Release (Diborane)

You have a tank containing 2,500 pounds of diborane gas. Assuming the total quantity in the tank is released over a 10-minute period, the release rate (QR), from Equation 3-1, is:

QR = 2,500 pounds/10 minutes = 250 pounds per minute

3.1.2 Releases of Toxic Gas in Enclosed Space

If a gas is released in an enclosure such as a building or shed, the release rate to the outside air may be lessened considerably. The dynamics of this type of release are complex; however, you may use the simplified method presented here to estimate an approximate release rate to the outside air from a release in an enclosed space. The mitigation factor (i.e., 55 percent) presented in this method assumes that the release occurs in a fully enclosed, non-airtight space that is directly adjacent to the outside air. If you are modeling a release in an interior room that is enclosed within a building, a smaller factor (i.e., more mitigation) may be appropriate. On the other hand, a larger factor (i.e., less mitigation) should be used for a space that has doors or windows that could be open during a release. If any of these special circumstances apply to your site, you may want to consider performing site-specific modeling to determine the appropriate amount of passive mitigation. In addition, you should not incorporate the passive mitigation effect of building enclosures into your modeling if you have reason to believe the enclosure would not withstand the force of the release or if the chemical is handled outside the building (e.g., moved from one building to another building).

For the worst case, assume as before that the largest quantity resulting from a pipe or vessel failure is released over a 10-minute period. Determine the unmitigated worst-case scenario release rate of the gas as the quantity released divided by 10 (Equation 3-1). The release rate from the building will be approximately 55 percent of the worst-case scenario release rate (see Appendix D, Section D.1.2 for the derivation of this factor). Estimate the mitigated release rate as follows:

$$QR = \frac{QS}{10} \times 0.55 \tag{3-2}$$

where: QR = Release rate (pounds per minute)
QS = Quantity released (pounds)
0.55 = Mitigation factor (discussed in Appendix D, Section D.1.2)

Example 2. Gas Release in Enclosure (Diborane)

Suppose the diborane gas from Example 1 is released inside a building at the rate of 250 pounds per minute. The mitigated release to the outside air from the building would be:

QR = 250 pounds/minute × 0.55 = 138 pounds per minute

3.1.3 Releases of Liquefied Refrigerated Toxic Gas in Diked Area

If you have a toxic gas that is liquefied by refrigeration alone, and it will be released into an area where it will be contained by dikes to form a pool more than one centimeter (0.033 foot) in depth, you may carry out the worst-case analysis assuming evaporation from a liquid pool at the boiling point of the liquid. If your gas liquefied by refrigeration would form a pool one centimeter (0.033 foot) or less in depth, use the methods described in Section 3.1.1 or 3.1.2 above for the analysis. For a release in a diked area, first compare the diked area to the maximum area of the pool that could be formed. You can use Equation 3-6 in Section 3.2.3 to estimate the maximum size of the pool. Density factors (DF), needed for Equation 3-6, for toxic gases at their boiling points are listed in Exhibit B-1 of Appendix B. If the pool formed by the released liquid would be smaller than the diked area, assume a 10-minute gaseous release, and estimate the release rate as described in Section 3.1.1. If the dikes prevent the liquid from spreading out to form a pool of maximum size (one centimeter in depth), you may use the method described in Section 3.2.3 for mitigated liquid releases to estimate a release rate from a pool at the boiling point of the released substance. Use Equation 3-8 in Section 3.2.3 for the release rate. The Liquid Factor Boiling (LFB) for each toxic gas, needed to use Equation 3-8, is listed in Exhibit B-1 of Appendix B. See the example release rate estimation on the next page.

After you have estimated the release rate, estimate the duration of the vapor release from the pool (the time it will take for the pool to evaporate completely) by dividing the total quantity spilled by the release rate. You need to know the duration of release to choose the appropriate reference table of distances to estimate the consequence distance, as discussed in Section 4. (You do not need to consider the duration of the release for chlorine or sulfur dioxide, liquefied by refrigeration alone. Only one reference table of distances is provided for worst-case releases of each of these substances, and these tables may be used regardless of the release duration. The principal reason for making no distinction between 10-minute and longer releases for the chemical-specific tables is that the differences between the two are small relative to the uncertainties that have been identified.)

Example 3. Mitigated Release of Gases Liquefied by Refrigeration (Chlorine)

You have a refrigerated tank containing 50,000 pounds of liquid chlorine at ambient pressure. A diked area around the chlorine tank of 275 square feet is sufficient to hold all of the spilled liquid chlorine. Once the liquid spills into the dike, it is then assumed to evaporate at its boiling point (-29 °F). The evaporation rate at the boiling point is determined from Equation 3-8. For the calculation, wind speed is assumed to be 1.5 meters per second and the wind speed factor is 1.4, LFB for chlorine (from Exhibit B-1) is 0.19, and A is 275 square feet. The release rate is:

$$QR = 1.4 \times 0.19 \times 275 = 73 \text{ pounds per minute}$$

The duration of the release does not need to be considered for chlorine.

3.2 Release Rates for Toxic Liquids

<u>In Section 3.2</u>

- 3.2.1 Method to estimate the quantity of toxic liquid that could be released from a broken pipe.
- 3.2.2 Method to estimate the release rate of a toxic liquid evaporating from a pool with no mitigation (no dikes or enclosures), including:
  - -- Releases at ambient temperature (25 °C),
  - -- Releases at elevated temperature, and
  - -- Estimation of the duration of the release.
- 3.2.3 Method to estimate the release rate of a toxic liquid evaporating from a pool with passive mitigation, including:
  - -- Releases in diked areas,
  - -- Releases into other types of containment, and
  - -- Releases into buildings.
- 3.2.4 Estimation of release rates for mixtures containing toxic liquids.
- 3.2.5 Method to correct the estimated release rate for liquids released at temperatures between 25 °C and 50 °C.

For the worst-case analysis, the release rate to air for toxic liquids is assumed to be the rate of evaporation from the pool formed by the released liquid. This section provides methods to estimate the evaporation rate. Assume the total quantity in a vessel or the maximum quantity from pipes is released into the pool. Passive mitigation measures (e.g., dikes) may be considered in determining the area of the pool and the release rate. To estimate the consequence distance using this guidance, you must estimate how long it will take for the pool to evaporate (the duration of the release), as well as the release rate, as discussed below.

The rule (40 CFR 68.22(g)) requires you to assume that liquids (other than gases liquefied by refrigeration) are released at the highest maximum daily temperature for the previous three years or at process temperature, whichever is higher. This chapter provides methods to estimate the release rate at 25 °C (77 °F) or at the boiling point, and also provides a method to correct the release rate at 25 °C for releases at temperatures between 25 °C and 50 °C.

The calculation methods provided in this section apply to substances that are liquids under ambient conditions or gases liquefied by refrigeration alone that are released to form pools deeper than one centimeter (see Section 3.1.3 above). You must treat gases liquefied under other conditions (under pressure or a combination of pressure and refrigeration) or gases liquefied by refrigeration alone that would form pools one centimeter or less in depth upon release as gas rather than liquid releases (see Sections 3.1.1 and 3.1.2 above).

3.2.1 Releases of Toxic Liquids from Pipes

To consider a liquid release from a broken pipe, estimate the maximum quantity that could be released assuming that the pipe is full of liquid. To estimate the quantity in the pipe, you need to know the length of the pipe (in feet) and cross-sectional area of the pipe (in square feet). Note also that liquid may be released from both directions at a pipe shear (both in the direction of operational flow and the reverse direction, depending on the location of the shear). Therefore, the length would be the full length of pipe carrying the liquid on the facility grounds. Then, the volume of the liquid in the pipe (in cubic feet) is the length of the pipe times the cross-sectional area. The quantity in the pipe (in pounds) is the volume divided by the Density Factor (DF) times 0.033. (DF values are listed in Appendix B, Exhibit B-2. Density in pounds per cubic foot is equal to 1/(DF times 0.033).) Assume the estimated quantity (in pounds) is released into a pool and use the method and equations described below in Section 3.2.2 (unmitigated releases) or 3.2.3 (releases with passive mitigation) to determine the evaporation rate of the liquid from the pool.

3.2.2 Unmitigated Releases of Toxic Liquids

If no passive mitigation measures are in place, the liquid is assumed to form a pool one centimeter (0.39 inch or 0.033 foot) deep instantaneously. You may calculate the release rate to air from the pool (the evaporation rate) as discussed below for releases at ambient or elevated temperature.

Ambient Temperature

If the liquid is always at ambient temperature, find the Liquid Factor Ambient (LFA) and the Density Factor (DF) in Exhibit B-2 of Appendix B. The LFA and DF apply to liquids at 25 °C; if your ambient temperature is between 25 °C and 50 °C, you may use the method described here and then apply a Temperature Correction Factor (TCF), as discussed in Section 3.2.5 below, to correct the calculated release rate. Calculate the release rate of the liquid at 25 °C from the following equation:

$$QR = QS \times 1.4 \times LFA \times DF \qquad (3\text{-}3)$$

where:	QR	=	Release rate (pounds per minute)
	QS	=	Quantity released (pounds)
	1.4	=	Wind speed factor = $1.5^{0.78}$, where 1.5 meters per second (3.4 miles per hour) is the wind speed for the worst case
	LFA	=	Liquid Factor Ambient
	DF	=	Density Factor

Example 4. Unmitigated Liquid Release at Ambient Temperature (Acrylonitrile)

You have a tank containing 20,000 pounds of acrylonitrile at ambient temperature. The total quantity in the tank is spilled onto the ground in an undiked area, forming a pool. Assume the pool spreads out to a depth of one centimeter. The release rate from the pool (QR) is calculated from Equation 3-3. For the calculation, the wind speed is assumed to be 1.5 meters per second and the wind speed factor is 1.4. From Exhibit B-2, Appendix B, LFA for acrylonitrile is 0.018 and DF is 0.61. Then:

$$QR = 20{,}000 \times 1.4 \times 0.018 \times 0.61 = 307 \text{ pounds per minute}$$

The duration of the release (from Equation 3-5) would be:

$$t = 20{,}000 \text{ pounds}/307 \text{ pounds per minute} = 65 \text{ minutes}$$

Elevated Temperature

If the liquid is at an elevated temperature (above 50 °C or at or close to the boiling point), find the Liquid Factor Boiling (LFB) and the Density Factor (DF) in Exhibit B-2 of Appendix B (see Appendix D, Section D.2.2, for the derivation of these factors). For temperatures up to 50 °C, you may use the method above for ambient temperature and apply the Temperature Correction Factors, as discussed in Section 3.2.5. If the temperature is above 50 °C, or the liquid is at or close to its boiling point, or no Temperature Correction Factors are available for your liquid, calculate the release rate of the liquid from the following equation:

$$QR = QS \times 1.4 \times LFB \times DF \qquad (3\text{-}4)$$

where:	QR	=	Release rate (pounds per minute)
	QS	=	Quantity released (pounds)
	1.4	=	Wind speed factor = $1.5^{0.78}$, where 1.5 meters per second (3.4 miles per hour) is the wind speed for the worst case
	LFB	=	Liquid Factor Boiling
	DF	=	Density Factor

Example 5. Unmitigated Release at Elevated Temperature (Acrylonitrile)

You have a tank containing 20,000 pounds of acrylonitrile at an elevated temperature. The total quantity in the tank is spilled onto the ground in an undiked area, forming a pool. Assume the pool spreads out to a depth of one centimeter. The release rate from the pool is calculated from Equation 3-4. For the calculation, the wind speed factor for 1.5 meters per second is 1.4. From Exhibit B-2, Appendix B, LFB for acrylonitrile is 0.11 and DF is 0.61. Then:

$$QR = 20{,}000 \times 1.4 \times 0.11 \times 0.61 = 1{,}880 \text{ pounds per minute}$$

The duration of the release (from Equation 3-5) would be:

$$t = 20{,}000 \text{ pounds}/1880 \text{ pounds per minute} = 11 \text{ minutes}$$

Duration of Release

After you have estimated a release rate as described above, determine the duration of the vapor release from the pool (the time it will take for the liquid pool to evaporate completely). If you calculate a corrected release rate for liquids above 25 °C, use the corrected release rate, estimated as discussed in Section 3.2.5 below, to estimate the release duration. To estimate the time in minutes, divide the total quantity released (in pounds) by the release rate (in pounds per minute) as follows:

$$t = \frac{QS}{QR} \tag{3-5}$$

where:

t	=	Duration of the release (minutes)
QR	=	Release rate (pounds per minute) (use release rate corrected for temperature, QR_{cr}, if appropriate)
QS	=	Quantity released (pounds)

You will use the duration of the vapor release from the pool to decide which table is appropriate for estimating distance, as discussed in Chapter 4 below.

3.2.3 Releases of Toxic Liquids with Passive Mitigation

Diked Areas

If the toxic liquid will be released into an area where it will be contained by dikes, compare the diked area to the maximum area of the pool that could be formed; the smaller of the two areas should be used in determination of the evaporation rate. The maximum area of the pool (assuming a depth of one centimeter) is:

$$A = QS \times DF \quad (3\text{-}6)$$

where: A = Maximum area of pool (square feet) for depth of one centimeter
QS = Quantity released (pounds)
DF = Density Factor (listed in Exhibit B-2, Appendix B)

Maximum Area Smaller than Diked Area. If the maximum area of the pool is smaller than the diked area, calculate the release rate as described for "no mitigation" above.

Diked Area Smaller than Maximum Area. If the diked area is smaller than the maximum pool area, go to Exhibit B-2 in Appendix B to find the Liquid Factor Ambient (LFA), if the liquid is at ambient temperature, or the Liquid Factor Boiling (LFB), if the liquid is at an elevated temperature. For liquids at temperatures between 25 °C and 50 °C, you may use the method described here and then apply a Temperature Correction Factor (TCF), as discussed in Section 3.2.5 below, to correct the calculated release rate. For gases liquefied by refrigeration alone, use LFB from Exhibit B-1. Calculate the release rate from the diked area as follows for liquids at ambient temperature:

$$QR = 1.4 \times LFA \times A \quad (3\text{-}7)$$

or, for liquids at elevated temperature or for gases liquefied by refrigeration alone:

$$QR = 1.4 \times LFB \times A \quad (3\text{-}8)$$

where: QR = Release rate (pounds per minute)
1.4 = Wind speed factor = $1.5^{0.78}$, where 1.5 meters per second (3.4 miles per hour) is the wind speed for the worst case
LFA = Liquid Factor Ambient (listed in Exhibit B-2, Appendix B)
LFB = Liquid Factor Boiling (listed in Exhibit B-1 (for liquefied gases) or B-2 (for liquids), Appendix B)
A = Diked area (square feet)

Potential Overflow of Diked Area. In case of a large liquid spill, you also need to consider whether the liquid could overflow the diked area. Follow these steps:

- Determine the volume of the diked area in cubic feet from surface area times depth or length times width times depth (in feet).
- Determine the volume of liquid spilled in cubic feet from QS × DF × 0.033 (DF × 0.033 is equal to 1/density in pounds per cubic foot).
- Compare the volume of the diked area to the volume of liquid spilled. If the volume of liquid is greater than the volume of the diked area:
  - -- Subtract the volume of the diked area from the total volume spilled to determine the volume that might overflow the diked area.

-- Estimate the maximum size of the pool formed by the overflowing liquid (in square feet) by dividing the overflow volume (in cubic feet) by 0.033 (the depth of the pool in feet).

-- Add the surface area of the diked area and the area of the pool formed by the overflow to estimate the total pool area (A).

-- Estimate the evaporation rate from Equation 3-7 or 3-8 above.

After you have estimated the release rate, estimate the duration of the vapor release from the pool by dividing the total quantity spilled by the release rate (Equation 3-5 above).

Example 6. Mitigated Liquid Release at Ambient Temperature (Bromine)

You have a tank containing 20,000 pounds of bromine at an ambient temperature of 25 °C. Assume that the total quantity in the tank is spilled into a square diked area 10 feet by 10 feet (area 100 square feet). The dike walls are four feet high. The area (A) that would be covered to a depth of 0.033 feet (one centimeter) by the spilled liquid is given by Equation 3-6 as the quantity released (QS) times the Density Factor (DF). From Exhibit B-2, Appendix B, DF for bromine is 0.16. Then:

$$A = 20{,}000 \times 0.16, \text{ or } 3{,}200 \text{ square feet}$$

The diked area is smaller than the maximum pool area. The volume of bromine spilled is 20,000 × 0.16 × 0.033, or 106 cubic feet. The spilled liquid would fill the diked area to a depth of a little more than one foot, well below the top of the wall. You use the diked area to determine the evaporation rate from Equation 3-7. For the calculation, wind speed is 1.5 meters per second, the wind speed factor is 1.4, LFA for bromine (from Exhibit B-2) is 0.073, and A is 100 square feet. The release rate is:

$$QR = 1.4 \times 0.073 \times 100 = 10 \text{ pounds per minute}$$

The maximum duration of the release would be:

$$t = 20{,}000 \text{ pounds} / 10 \text{ pounds per minute} = 2{,}000 \text{ minutes}$$

Other Containment

If the toxic liquid will be contained by other means (e.g., enclosed catch basins or trenches), consider the total quantity that could be spilled and estimate the surface area of the released liquid that potentially would be exposed to the air. Look at the dimensions of trenches or other areas where spilled liquids would be exposed to the air to determine the surface area of pools that could be formed. Use the instructions above to estimate a release rate from the total surface area.

Releases into Buildings

If the toxic liquid is released inside a building, compare the area of the pool that would be formed (depending upon floor space or passive mitigation) to the maximum area of the pool that could be formed (if the liquid is not contained); the smaller of the two areas should be used in determining the evaporation rate. The maximum area of the pool is determined as described above for releases into diked areas, using Equation 3-6. If the toxic liquid would spread to cover the building floor, you determine the area of the building floor as:

$$A = L \times W \tag{3-9}$$

where:	A	=	Area (square feet)
	L	=	Length (feet)
	W	=	Width (feet)

If there are obstacles such as dikes inside the building, determine the size of the pool that would be formed based on the area defined by the dikes or other obstacles.

The evaporation rate is then determined for a worst-case scenario (i.e., wind speed is 1.5 meters per second (3.4 miles per hour)), using Equation 3-3 or 3-4, if the liquid spreads to its maximum area, or Equation 3-7 or 3-8, if the pool area is smaller than the maximum. The maximum rate of evaporated liquid exiting the building is taken to be 10 percent of the calculated worst-case scenario evaporation rate (see Appendix D, Section D.2.4 for the derivation of this factor), as follows:

$$QR_B = 0.1 \times QR \tag{3-10}$$

where:	QR_B	=	Release rate from building
	QR	=	Release rate from pool, estimated as discussed above
	0.1	=	Mitigation factor, discussed in Appendix D, Section D.2.4

Note that the mitigation factor (i.e., 0.1) presented in this method assumes that the release occurs in a fully enclosed, non-airtight space that is directly adjacent to the outside air. It may not apply to a release in an interior room that is enclosed within a building, or to a space that has doors or windows that could be open during a release. In such cases, you may want to consider performing site-specific modeling to determine the appropriate amount of passive mitigation.

Example 7. Liquid Release Inside Building (Bromine)

Suppose that your tank of bromine from Example 6 is contained inside a storage shed 10 feet by 10 feet (area 100 square feet). There are no dikes inside the shed. From Example 6, you see that the area covered by the bromine in an unenclosed space would be 3,200 square feet. The building area is smaller than the maximum pool area; therefore, the building floor area should be used to determine the evaporation rate from Equation 3-7. For the calculation, first determine the worst-case scenario evaporation rate:

$$QR = 1.4 \times 0.073 \times 100 = 10 \text{ pounds per minute}$$

The release rate to the outside air of the evaporated liquid leaving the building would then be:

$$QR_B = 0.1 \times 10 \text{ pounds per minute} = 1 \text{ pound per minute}$$

3.2.4 Mixtures Containing Toxic Liquids

Mixtures containing regulated toxic substances do not have to be considered if the concentration of the regulated substance in the mixture is below one percent by weight or if you can demonstrate that the partial vapor pressure of the regulated substances in the mixture is below 10 millimeters of mercury (mm Hg). Regulated substances present as by-products or impurities would need to be considered if they are present in concentrations of one percent or greater in quantities above their thresholds, and their partial vapor pressures are 10 mm Hg or higher. In case of a spill of a liquid mixture containing a regulated toxic substance with partial vapor pressure of 10 mm Hg or higher (with the exception of common water solutions, discussed in the next section), you have several options for estimating a release rate:

- Carry out the analysis as described above in Sections 3.2.2 or 3.2.3 using the quantity of the regulated substance in the mixture and the liquid factor (LFA or LFB) and density factor for the regulated substance in pure form. This is a simple approach that likely will give conservative results.

- If you know the partial pressure of the regulated substance in the mixture, you may estimate a more realistic evaporation rate. An equation for the evaporation rate is given at the end of Section B.2 in Appendix B.

  -- In this case, estimate a pool size for the entire quantity of the mixture, for an unmitigated release. If you know the density of the mixture, you may use it in estimating the pool size; otherwise, you may assume the density is the same as the pure regulated substance (in most cases, this assumption is unlikely to have a large effect on the results).

- You may estimate the partial pressure of the regulated substance in the mixture by the method described in Section B.2 in Appendix B and use the equation presented there to estimate an evaporation rate. This equation is appropriate to mixtures and solutions in which the components do not interact with each other. It is probably inappropriate for most water solutions. It is likely to overestimate the partial vapor pressure of regulated substances in water solutions in which hydrogen bonding may occur (e.g., solutions of acids or alcohols). As discussed above, use the pool size for the entire quantity of the mixture for an unmitigated release.

Example 8. Mixture Containing Toxic Liquid (Acrylonitrile)

You have a tank containing 50,000 pounds of a mixture of acrylonitrile (a regulated substance) and N,N-dimethylformamide (not regulated). The weight of each of the components of the mixture is known (acrylonitrile = 20,000 pounds; N,N-dimethylformamide = 30,000 pounds.) The molecular weight of acrylonitrile, from Exhibit B-2, is 53.06, and the molecular weight of N,N-dimethylformamide is 73.09. Using Equation B-3, Appendix B, calculate the mole fraction of acrylonitrile in the solution as follows:

$$X_r = \frac{(20{,}000/53.06)}{(20{,}000/53.06) + (30{,}000/73.09)}$$

$$X_r = \frac{377}{377 + 410}$$

$$X_r = 0.48$$

Estimate the partial vapor pressure of acrylonitrile using Equation B-4 as follows (using the vapor pressure of acrylonitrile in pure form at 25 C, 108 mm Hg, from Exhibit B-2, Appendix B):

$$VP_m = 0.48 \times 108 = 51.8 \text{ mm Hg}$$

Before calculating evaporation rate for acrylonitrile in the mixture, you must determine the surface area of the pool formed by the entire quantity of the mixture, using Equation 3-6. The quantity released is 50,000 pounds and the Density Factor for acrylonitrile is 0.61 in Exhibit B-2; therefore:

$$A = 50{,}000 \times 0.61 = 30{,}500 \text{ square feet}$$

Now calculate the evaporation rate for acrylonitrile in the mixture from Equation B-5 using the VP_m and A calculated above:

$$QR = \frac{0.0035 \times 1.0 \times (53.06)^{2/3} \times 30{,}500 \times 51.8}{298}$$

$$QR = 262 \text{ pounds per minute}$$

3.2.5 Release Rate Correction for Toxic Liquids Released at Temperatures Between 25 °C and 50 °C

If your liquid is at a temperature between 25 °C (77 °F) and 50 °C (122 °F), you must use the higher temperature for the offsite consequence analysis. You may correct the release rate calculated for a pool at 25

°C to estimate from a pool at the higher temperature using Temperature Correction Factors (TCF) provided in Appendix B, Exhibit B-4. Calculate a corrected release rate as follows:

- Calculate the release rate (QR) of the liquid at 25 °C (77 °F) as described in Section 3.2.2 (for unmitigated releases) or 3.2.3 (for releases with passive mitigation).

- From Exhibit B-4 in Appendix B:

  -- Find your liquid in the left-hand column of the table.

  -- Find the temperature closest to your temperature at the top of the table. If your temperature is at the midpoint between two temperatures, go to the higher temperature; otherwise go to the closest temperature (higher or lower than your temperature).

  -- Find the TCF for your liquid in the column for the appropriate temperature.

- Estimate a corrected release rate (QR_C) by multiplying the estimated release rate by the TCF; i.e.,

$$QR_C = QR \times TCF \tag{3-11}$$

where: QR_C = Corrected release rate
QR = Release rate calculated for 25 °C
TCF = Temperature Correction Factor (from Exhibit B-4, Appendix B)

The derivation of the Temperature Correction Factors is discussed in Appendix D, Section D.2.2. If you have vapor pressure-temperature data for a liquid not covered in Exhibit B-4, you may correct the evaporation rate using the method presented in Section D.2.2.

Example 9. Liquid Release at Ambient Temperature Between 25 °C and 50 °C (Bromine)

Assume the tank containing 20,000 pounds of bromine, from Example 6, is at an ambient temperature of 35 °C (95 °F). As in Example 6, the total quantity in the tank is spilled into a diked enclosure that completely contains the spill. The surface area is 100 square feet. In Example 6, the release rate (QR) at 25 °C was calculated from Equation 3-7 to be 10 pounds per minute. To adjust the release rate for the temperature of 35 °C, you find the Temperature Correction Factor (TCF) for bromine at 35 °C from Exhibit B-4 in Appendix B. The TCF at this temperature is 1.5; the corrected release rate (QR_C) at 35 °C, from Equation 3-11, is

$$QR_C = 10 \times 1.5 = 15 \text{ pounds per minute}$$

The duration of the release (from Equation 3-5) would be:

$$t = 20{,}000 \text{ pounds}/15 \text{ pounds per minute} = 1{,}300 \text{ minutes}$$

3.3 Release Rates for Common Water Solutions of Toxic Substances and for Oleum

<u>In Section 3.3</u>

- Methods to estimate the release rates for several common water solutions and for oleum, including:
  - -- Evaporation from pools with no mitigation (see 3.2.2),
  - -- Evaporation from pools with dikes (see 3.2.3),
  - -- Releases at elevated temperatures of solutions of gases, and
  - -- Releases at elevated temperatures of solutions of liquids.

This section presents a simple method of estimating the release rate from spills of water solutions of several substances. Oleum (a solution of sulfur trioxide in sulfuric acid) also is discussed in this section.

The vapor pressure and evaporation rate of a substance in solution depends on its concentration in the solution. If a concentrated water solution containing a volatile toxic substance is spilled, the toxic substance initially will evaporate more quickly than water from the spilled solution, and the vapor pressure and evaporation rate will decrease as the concentration of the toxic substance in the solution decreases. At much lower concentrations, water may evaporate more quickly than the toxic substance. There is one concentration at which the composition of the solution does not change as evaporation occurs. For most situations of interest, the concentration exceeds this concentration, and the toxic substance evaporates more quickly than water.

For estimating release rates from solutions, this guidance lists liquid factors (ambient) for several common water solutions at several concentrations that take into account the decrease in evaporation rate with decreasing concentration. Exhibit B-3 in Appendix B provides LFA and DF values for several concentrations of ammonia, formaldehyde, hydrochloric acid, hydrofluoric acid, and nitric acid in water solution. Factors for oleum are also included in the exhibit. Chlorine dioxide also may be found in water solutions; however, solutions of chlorine dioxide commonly are below one percent concentration. Solutions below one percent concentration do not have to be considered. Chlorine dioxide, therefore, is not included in Exhibit B-3. These factors may be used to estimate an average release rate for the listed substances from a pool formed by a spill of solution. Liquid factors are provided for two different wind speeds, because the wind speed affects the rate of evaporation.

For the worst case, use the factor for a wind speed of 1.5 meters per second (3.4 miles per hour). You need to consider only the first 10 minutes of the release for solutions under ambient conditions in estimating the consequence distance, because the toxic component in a solution evaporates fastest during the first few minutes of a spill, when its concentration is highest. Modeling indicates that analysis considering the first 10 minutes of the release gives a good approximation of the overall consequences of the release. Although the toxic substance will continue to evaporate from the pool after 10 minutes, the rate of evaporation is so much lower that it can safely be ignored in estimating the consequence distance. (See Appendix D, Section D.2.3, for more information.) Estimate release rates as follows:

Ambient Temperature

- Unmitigated. If no passive mitigation measures are in place, and the solution is at ambient temperature, find the LFA at 1.5 meters per second (3.4 miles per hour) and DF for the solution in Appendix B, Exhibit B-3. Follow the instructions for liquids presented in Section 3.2.2 above to estimate the release rate of the listed substance in solution. Use the total quantity of the solution as the quantity released (QS) in carrying out the calculation of release rate.

- Mitigated. If passive mitigation is in place, and the solution is at ambient temperature, find the LFA at 1.5 meters per second (3.4 miles per hour) in Appendix B, Exhibit B-3, and follow the instructions for liquids in Section 3.2.3 above. Use the total quantity of the solution to estimate the maximum pool area for comparison with the diked area.

Example 10. Evaporation Rate for Water Solution at Ambient Temperature (Hydrochloric Acid)

You have a tank containing 50,000 pounds of 37 percent hydrochloric acid solution, at ambient temperature. For the worst-case analysis, you assume the entire contents of the tank is released, forming a pool. The release occurs in a diked area of 9,000 square feet.

From Exhibit B-3, Appendix B, the Density Factor (DF) for 37 percent hydrochloric acid is 0.42. From Equation 3-6, the maximum area of the pool would be 50,000 times 0.42, or 21,000 square feet. The diked area is smaller; therefore, the diked area should be used in the evaporation rate (release rate) calculation, using Equation 3-7.

For the calculation using Equation 3-7, you need the pool area (9,000 square feet) and the Liquid Factor Ambient (LFA) for 37 percent hydrochloric acid; you assume a wind speed of 1.5 meters per second, so the wind speed factor is 1.4. From Exhibit B-3, Appendix B, the LFA is 0.0085. From Equation 3-7, the release rate (QR) of hydrogen chloride from the pool is:

$$QR = 1.4 \times 9{,}000 \times 0.0085 = 107 \text{ pounds per minute}$$

You do not need to consider the duration of the release, because only the first ten minutes are considered.

Elevated Temperature

- Known Vapor Pressure. If the solution is at an elevated temperature, the vapor pressure of the regulated substance and its release rate from the solution will be much higher. This guidance does not include temperature correction factors for evaporation rates of regulated substances from solutions. If you know the partial vapor pressure of the toxic substance in solution at the relevant temperature, you can carry out the calculation of the release rate using the equations in Appendix D, Sections D.2.1 and D.2.2. As for releases of solutions at ambient temperature, you only need to consider the first 10 minutes of the release, because the evaporation rate of the toxic substance from the solution will decrease rapidly as its concentration decreases.

- Unknown Vapor Pressure. If you do not know the vapor pressure of the substance in solution, as a conservative approach for the worst-case analysis, use the appropriate instructions, as follows:

  -- *Solutions containing substances that are gases under ambient conditions*. The list of regulated substances includes several substances that, in their pure form, are gases under ambient conditions, but that may commonly be found in water solutions. These substances include ammonia, formaldehyde, hydrogen chloride, and hydrogen fluoride. For a release of a solution of ammonia, formaldehyde, hydrochloric acid, or hydrofluoric acid above ambient temperature, if you do not have vapor pressure data for the temperature of interest or prefer a simpler method, assume the quantity of the pure substance in the solution is released as a gas over 10 minutes, as discussed in Section 3.1 above. You may determine the amount of pure substance in the solution from the concentration (e.g., a solution of 37 percent hydrochloric acid by weight would contain a quantity of hydrogen chloride equal to 0.37 times the total weight of the solution).

Example 11. Evaporation Rate for Gas in Water Solution at Elevated Temperature (Hydrochloric Acid)

You have 50,000 pounds of 37 percent hydrochloric acid solution in a high-temperature process. For the worst-case analysis, you assume the entire contents of the process vessel is released. In this case, because the solution is at an elevated temperature, you consider the release of gaseous hydrogen chloride from the hot solution.

The solution would contain 50,000 × 0.37 pounds of hydrogen chloride, or 18,500 pounds. You assume the entire 18,500 pounds is released over 10 minutes. From Equation 3-1, the release rate is 18,500 divided by 10, or 1,850 pounds per minute.

-- *Liquids in solution*. If you have vapor pressure data for the liquid in solution (including nitric acid in water solution and sulfur trioxide in oleum) at the temperature of interest, you may use that data to estimate the release rate, as discussed above. You only need to consider the first 10 minutes of the evaporation.

For a release of nitric acid solution at a temperature above ambient, if you do not have vapor pressure data or prefer to use this simpler method, determine the quantity of pure nitric acid in the solution from the concentration. Assume the quantity of pure nitric acid is released at an elevated temperature and estimate a release rate as discussed in Section 3.2 above, using the LFB. For temperatures between 25 °C and 50 °C, you may use the LFA and the temperature correction factors for the pure substance, as described in Section 3.2.5. You do not need to estimate the duration of the release, because you only consider the first 10 minutes.

Similarly, for a release of oleum at an elevated temperature, determine the quantity of free sulfur trioxide in the oleum from the concentration and assume the sulfur trioxide is released at an elevated temperature. Use the LFB or the LFA and temperature correction factors for sulfur trioxide to estimate a release rate as discussed in Section 3.2. You only need to consider the first 10 minutes of the release in your analysis.

For a spill of liquid in solution into a diked area, you would need to consider the total quantity of <u>solution</u> in determining whether the liquid could overflow the diked area (see the steps in Section 3.2.3). If you find that the liquid could overflow the dikes, you would need to consider both the quantity of pure substance remaining inside the diked area and the quantity of pure substance spilled outside the diked area in carrying out the release rate analysis as discussed in Section 3.2.3.

Example 12. Evaporation Rate for Liquid in Water Solution at Elevated Temperature (Nitric Acid)

You have 18,000 pounds of 90 percent nitric acid solution in a high temperature process. The solution would contain 18,000 x 0.90 pounds of nitric acid, or 16,200 pounds. You assume 16,200 pounds of pure nitric acid is released at an elevated temperature.

For the calculation using Equation 3-4, you need the quantity released (16,200); the Liquid Factor Boiling (LFB) for nitric acid (0.12 from Exhibit B-2); the Density Factor (DF) for nitric acid (0.32 from Exhibit B-2); and you assume a wind speed of 1.5 meter per second, so the wind speed factor is 1.4. From Equation 3-4, the release rate (QR) of hot nitric acid is:

$$QR = 16{,}200 \times 1.4 \times 0.12 \times 0.32 = 870 \text{ pounds per minute}$$

You do not need to estimate the duration of release, because you only consider the first 10 minutes.

4 ESTIMATION OF WORST-CASE DISTANCE TO TOXIC ENDPOINT

<u>In Chapter 4</u>

- Reference tables of distances for worst-case releases, including:
  - -- Generic reference tables (Exhibit 2), and
  - -- Chemical-specific reference tables (Exhibit 3).
- Considerations include:
  - -- Gas density (neutrally buoyant or dense),
  - -- Duration of release (10 minutes or 60 minutes),
  - -- Topography (rural or urban).

This guidance provides reference tables giving worst-case distances for neutrally buoyant gases and vapors and for dense gases and vapors for both rural (open) and urban (obstructed) areas. This chapter describes these reference tables and gives instructions to help you choose the appropriate table to estimate consequence distances for the worst-case analysis.

Neutrally buoyant gases and vapors have approximately the same density as air, and dense gases and vapors are heavier than air. Neutrally buoyant and dense gases are dispersed in different ways when they are released; therefore, modeling was carried out to develop separate reference tables. These generic reference tables can be used to estimate distances using the specified toxic endpoint for each substance and the estimated release rate to air. In addition to the generic tables, chemical-specific reference tables are provided for ammonia, chlorine, and sulfur dioxide. These chemical-specific tables were developed based on modeling carried out for industry-specific guidance documents. All the tables were developed assuming a wind speed of 1.5 meters per second (3.4 miles per hour) and F stability. To use the reference tables, you need the worst-case release rates estimated as described in the previous sections. For liquid pool evaporation, you also need the duration of the release. In addition, to use the generic tables, you will need to determine the appropriate toxic endpoint and whether the gas or vapor is neutrally buoyant or dense, using the exhibits in Appendix B. You may interpolate between entries in the reference tables.

Generic reference tables are provided for both 10-minute releases and 60-minute releases. You should use the tables for 10-minute releases if the duration of your release is 10 minutes or less; use the tables for 60-minute releases if the duration of your release is more than 10 minutes. For the worst-case analysis, all releases of toxic gases are assumed to last for 10 minutes. You need to consider the estimated duration of the release (from Equation 3-5) for evaporation of pools of toxic liquids. For evaporation of water solutions of toxic liquids or of oleum, you should always use the tables for 10-minute releases.

The generic reference tables of distances (Reference Tables 1-8), which should be used for substances other than ammonia, chlorine, and sulfur dioxide, are found at the end of Chapter 5. The generic tables and the conditions for which each table are applicable are described in Exhibit 2. Chemical-specific reference tables of distances (Reference Tables 9-12) follow the generic reference tables at the end of Chapter 5. Each of these chemical-specific tables includes distances for both rural and urban topography. These tables are described in Exhibit 3.

Remember that these reference tables provide only rough estimates, not accurate predictions, of the distances that might be reached under worst-case conditions. In particular, although the distances in the tables are as great as 25 miles, you should bear in mind that the larger distances (more than six to ten miles) are very uncertain.

To use the reference tables of distances, follow these steps:

For Regulated Toxic Substances Other than Ammonia, Chlorine, and Sulfur Dioxide

- Find the toxic endpoint for the substance in Appendix B (Exhibit B-1 for toxic gases or Exhibit B-2 for toxic liquids).

- Determine whether the table for neutrally buoyant or dense gases and vapors is appropriate from Appendix B (Exhibit B-1 for toxic gases or Exhibit B-2 for toxic liquids). A toxic gas that is lighter than air may behave as a dense gas upon release if it is liquefied under pressure, because the released gas may be mixed with liquid droplets, or if it is cold. Consider the state of the released gas when you decide which table is appropriate.

- Determine whether the table for rural or urban conditions is appropriate.

  -- Use the rural table if your site is in an open area with few obstructions.

  -- Use the urban table if your site is in an urban or obstructed area. The urban tables are appropriate if there are many obstructions in the area, even if it is in a remote location, not in a city.

- Determine whether the 10-minute table or the 60-minute table is appropriate.

  -- Always use the 10-minute table for worst-case releases of toxic gases.

  -- Always use the 10-minute table for worst-case releases of common water solutions and oleum from evaporating pools, for both ambient and elevated temperatures.

  -- If you estimated the release duration for an evaporating toxic liquid pool to be 10 minutes or less, use the 10-minute table.

  -- If you estimated the release duration for an evaporating toxic liquid pool to be more than 10 minutes, use the 60-minute table.

Exhibit 2
Generic Reference Tables of Distances for Worst-case Scenarios

Applicable Conditions			Reference Table Number
Gas or Vapor Density	Topography	Release Duration (minutes)	
Neutrally buoyant	Rural	10	1
		60	2
	Urban	10	3
		60	4
Dense	Rural	10	5
		60	6
	Urban	10	7
		60	8

Exhibit 3
Chemical-Specific Reference Tables of Distances for Worst-case Scenarios

Substance	Applicable Conditions			Reference Table Number
	Gas or Vapor Density	Topography	Release Duration (minutes)	
Anhydrous ammonia liquefied under pressure	Dense	Rural, Urban	10	9
Non-liquefied ammonia, ammonia liquefied by refrigeration, or aqueous ammonia	Neutrally buoyant	Rural, Urban	10	10
Chlorine	Dense	Rural, Urban	10	11
Sulfur dioxide (anhydrous)	Dense	Rural, Urban	10	12

Neutrally Buoyant Gases or Vapors

- If Exhibit B-1 or B-2 indicates the gas or vapor should be considered neutrally buoyant, and other factors would not cause the gas or vapor to behave as a dense gas, divide the estimated release rate (pounds per minute) by the toxic endpoint (milligrams per liter).

- Find the range of release rate/toxic endpoint values that includes your calculated release rate/toxic endpoint in the first column of the appropriate table (Reference Table 1, 2, 3, or 4), then find the corresponding distance to the right (see Example 13 below).

Dense Gases or Vapors

- If Exhibit B-1 or B-2 or consideration of other relevant factors indicates the substance should be considered a dense gas or vapor (heavier than air), find the distance in the appropriate table (Reference Table 5, 6, 7, or 8) as follows;

  -- Find the toxic endpoint closest to that of the substance by reading across the top of the table. If the endpoint of the substance is halfway between two values on the table, choose the value on the table that is smaller (to the left). Otherwise, choose the closest value to the right or the left.

  -- Find the release rate closest to the release rate estimated for the substance at the left of the table. If the calculated release rate is halfway between two values on the table, choose the release rate that is larger (farther down on the table). Otherwise, choose the closest value (up or down on the table).

  -- Read across from the release rate and down from the endpoint to find the distance corresponding to the toxic endpoint and release rate for your substance.

For Ammonia, Chlorine, or Sulfur Dioxide

- Find the appropriate chemical-specific table for your substance (see the descriptions of Reference Tables 9-12 in Exhibit 3).

  -- If you have ammonia liquefied by refrigeration alone, you may use Reference Table 10, even if the duration of the release is greater than 10 minutes.

  -- If you have chlorine or sulfur dioxide liquefied by refrigeration alone, you may use the chemical-specific reference tables, even if the duration of the release is greater than 10 minutes.

- Determine whether rural or urban topography is applicable to your site.

  -- Use the rural column in the reference table if your site is in an open area with few obstructions.

- -- Use the urban column if your site is in an urban or obstructed area. The urban column is appropriate if there are many obstructions in the area, even if it is in a remote location, not in a city.

- Estimate the consequence distance as follows:

  - -- In the left-hand column of the table, find the release rate closest to your calculated release rate.

  - -- Read the corresponding distance from the appropriate column (urban or rural) to the right.

The development of Reference Tables 1-8 is discussed in Appendix D, Sections D.4.1 and D.4.2. The development of Reference Tables 9-12 is discussed in industry-specific risk management program guidance documents and a backup information document that are cited in Section D.4.3. If you think the results of the method presented here overstate the potential consequences of a worst-case release at your site, you may choose to use other methods or models that take additional site-specific factors into account.

Examples 14 and 15 below include the results of modeling using two other models, the Areal Locations of Hazardous Atmospheres (ALOHA) and the World Bank Hazards Analysis (WHAZAN) systems. These additional results are provided for comparison with the results of the methods presented in this guidance. The same modeling parameters were used as in the modeling carried out for development of the reference tables of distances. Appendix D, Section D.4.5, provides information on the modeling carried out with ALOHA and WHAZAN.

Example 13. Gas Release (Diborane)

In Example 1, you estimated a release rate for diborane gas of 250 pounds per minute. From Exhibit B-1, the toxic endpoint for diborane is 0.0011 mg/L, and the appropriate reference table for diborane is a neutrally buoyant gas table. Your facility and the surrounding area have many buildings, pieces of equipment, and other obstructions; therefore, you assume urban conditions. The appropriate reference table is Reference Table 3, for a 10-minute release of a neutrally buoyant gas in an urban area.

The release rate divided by toxic endpoint for this example is 250/0.0011 = 230,000.

From Reference Table 3, release rate divided by toxic endpoint falls between 221,000 and 264,000, corresponding to about 8.1 miles.

Example 14. Gas Release (Ethylene Oxide)

You have a tank containing 10,000 pounds of ethylene oxide, which is a gas under ambient conditions. Assuming the total quantity in the tank is released over a 10-minute period, the release rate (QR) from Equation 3-1 is:

QR = 10,000 pounds/10 minutes = 1,000 pounds per minute

From Exhibit B-1, the toxic endpoint for ethylene oxide is 0.09 mg/L, and the appropriate reference table is the dense gas table. Your facility is in an open, rural area with few obstructions; therefore, you use the table for rural areas.

Using Reference Table 5 for 10-minute releases of dense gases in rural areas, the toxic endpoint of 0.09 mg/L is closer to 0.1 than 0.075 mg/L. For a release rate of 1,000 pounds per minute, the distance to 0.1 mg/L is 3.6 miles.

Additional Modeling for Comparison

The ALOHA model gave a distance of 2.2 miles to the endpoint, using the same assumptions.

The WHAZAN model gave a distance of 2.7 miles to the endpoint, using the same assumptions and the dense cloud dispersion model.

Example 15. Liquid Evaporation from Pool (Acrylonitrile)

You estimated an evaporation rate of 307 pounds per minute for acrylonitrile from a pool formed by the release of 20,000 pounds into an undiked area (Example 4). You estimated the time for evaporation of the pool as 65 minutes. From Exhibit B-2, the toxic endpoint for acrylonitrile is 0.076 mg/L, and the appropriate reference table for a worst-case release of acrylonitrile is the dense gas table. Your facility is in an urban area. You use Reference Table 8 for 60-minute releases of dense gases in urban areas.

From Reference Table 8, the toxic endpoint closest to 0.076 mg/L is 0.075 mg/L, and the closest release rate to 307 pounds per minute is 250 pounds per minute. Using these values, the table gives a worst-case consequence distance of 2.9 miles.

Additional Modeling for Comparison

The ALOHA model gave a distance of 1.3 miles to the endpoint for a release rate of 307 pounds per minute, using the same assumptions.

The WHAZAN model gave a distance of 1.0 mile to the endpoint for a release rate of 307 pounds per minute, using the same assumptions and the dense cloud dispersion model.

5 ESTIMATION OF DISTANCE TO OVERPRESSURE ENDPOINT FOR FLAMMABLE SUBSTANCES

In Chapter 5

- Methods to estimate the worst-case consequence distances for releases of flammable substances.
  - -- 5.1 Vapor cloud explosions of flammable substances that are not mixed with other substances, and
  - -- 5.2 Vapor cloud explosions of flammable substances in mixtures.

For the worst-case scenario involving a release of flammable gases and volatile flammable liquids, you must assume that the total quantity of the flammable substance forms a vapor cloud within the upper and lower flammability limits and the cloud detonates. As a conservative worst-case assumption, if you use the method presented here, you must assume that 10 percent of the flammable vapor in the cloud participates in the explosion. You need to estimate the consequence distance to an overpressure level of 1 pound per square inch (psi) from the explosion of the vapor cloud. An overpressure of 1 psi may cause partial demolition of houses, which can result in serious injuries to people, and shattering of glass windows, which may cause skin laceration from flying glass.

This chapter presents a simple method for estimating the distance to the endpoint for a vapor cloud explosion of a regulated substance. The method presented here for analysis of vapor cloud explosions is based on a TNT-equivalent model. Other methods are available for analysis of vapor cloud explosions, including methods that consider site-specific conditions. You may use other methods for your worst-case analysis if you so choose, provided you assume the total quantity of flammable substance is in the cloud and use an endpoint of 1 psi. If you use a TNT-equivalent model, you must assume a yield factor of 10 percent. Appendix A includes references to documents and journal articles on vapor cloud explosions that may provide useful information on methods of analysis.

5.1 Flammable Substances Not in Mixtures

For the worst-case analysis of a regulated flammable substance that is not in a mixture with other substances, you may estimate the consequence distance for a given quantity of a regulated flammable substance using Reference Table 13. This table provides distances to 1 psi overpressure for vapor cloud explosions of quantities from 500 to 2,000,000 pounds. These distances were estimated by a TNT-equivalent model, Equation C-1 in Appendix C, Section C.1, using the worst-case assumptions described above and data provided in Exhibit C-1, Appendix C. If you prefer, you may calculate your worst-case consequence distance for flammable substances from the heat of combustion of the flammable substance and Equation C-1 or C-2.

Example 16. Vapor Cloud Explosion (Propane)

You have a tank containing 50,000 pounds of propane. From Reference Table 13, the distance to 1 psi overpressure is 0.3 miles for 50,000 pounds of propane.

Alternatively, you can calculate the distance to 1 psi using Equation C-2 from Appendix C:

$$D = 0.0081 \times [\,0.1 \times 50{,}000 \times (46{,}333/4{,}680)\,]^{1/3}$$

$$D = 0.3 \text{ miles}$$

5.2 Flammable Mixtures

If you have more than 10,000 pounds of a mixture of flammable substances that meets the criteria for listing under CAA section 112(r) (flash point below 22.8 °C (73 °F), boiling point below 37.8 °C (100 °F), National Fire Protection Association (NFPA) flammability hazard rating of 4), you may need to carry out a worst-case consequence analysis for the mixture. (If the mixture itself does not meet these criteria, it is not covered, and no analysis is required, even if the mixture contains one or more regulated substances.) You should carry out the analysis using the total quantity of all regulated flammable substance or substances in the mixture. Non-flammable components should not be included. However, if additional (non-regulated) flammable substances are present in the mixture, you should include them in the quantity used in the analysis.

For simplicity, you may carry out the worst-case analysis based on the predominant regulated flammable component of the mixture or a major component of the mixture with the highest heat of combustion if the whole vapor cloud consists of flammable substances (see Exhibit C-1, Appendix C for data on heat of combustion). Estimate the consequence distance from Reference Table 13 for the major component with the highest heat of combustion, assuming that the quantity in the cloud is the total quantity of the mixture. If you have a mixture in which the heats of combustion of the components do not differ significantly (e.g., a mixture of hydrocarbons), this method is likely to give reasonable results.

Alternatively, you may estimate the heat of combustion of the mixture from the heats of combustion of the components of the mixture using the method described in Appendix C, Section C.2, and then use Equation C-1 or C-2 in Appendix C to determine the vapor cloud explosion distance. This method may be appropriate if you have a mixture that includes components with significantly different heats of combustion (e.g., a mixture of hydrogen and hydrocarbons) that make up a significant portion of the mixture.

Examples 17 and 18 illustrate the two methods of analysis. In Example 17, the heat of combustion of the mixture is estimated, and the distance to the endpoint is calculated from Equation C-2. In Example 18, the component of the mixture with the highest heat of combustion is assumed to represent the entire mixture, and the distance to the endpoint is read from Reference Table 13. For the mixture of two hydrocarbons used in the example, the methods give very similar results.

Example 17. Estimating Heat of Combustion of Mixture for Vapor Cloud Explosion Analysis

You have a mixture of 8,000 pounds of ethylene (the reactant) and 2,000 pounds of isobutane (a catalyst carrier). To carry out the worst-case analysis, estimate the heat of combustion of the mixture from the heats of combustion of the components of the mixture. (Ethylene heat of combustion = 47,145 kilojoules per kilogram; isobutane heat of combustion = 45,576). Using Equation C-3, Appendix C:

$$HC_m = \left[\frac{(8{,}000/2.2)}{(10{,}000/2.2)} \times 47{,}145 \right] + \left[\frac{(2{,}000/2.2)}{(10{,}000/2.2)} \times 45{,}576 \right]$$

$$HC_m = (37{,}716) + (9{,}115)$$

$$HC_m = 46{,}831 \text{ kilojoules per kilogram}$$

Now use the calculated heat of combustion for the mixture in Equation C-2 to calculate the distance to 1 psi overpressure for vapor cloud explosion.

$$D = 0.0081 \times [\, 0.1 \times 10{,}000 \times (46{,}831/4{,}680) \,]$$

$$D = 0.2 \text{ miles}$$

Example 18. Vapor Cloud Explosion of Flammable Mixture (Ethylene and Isobutane)

You have 10,000 pounds of a mixture of ethylene (the reactant) and isobutane (a catalyst carrier). To carry out the worst-case analysis, assume the quantity in the cloud is the total quantity of the mixture. Use data for ethylene because it is the component with the highest heat of combustion. (Ethylene heat of combustion = 47,145 kilojoules per kilogram; isobutane heat of combustion = 45,576, from Exhibit C-1, Appendix C). From Reference Table 13, the distance to 1 psi overpressure is 0.2 miles for 10,000 pounds of ethylene; this distance would also apply to the 10,000-pound mixture of ethylene and isobutane.

Reference Table 1

Neutrally Buoyant Plume Distances to Toxic Endpoint for Release Rate Divided by Endpoint
10-Minute Release, Rural Conditions, F Stability, Wind Speed 1.5 Meters per Second

Release Rate/Endpoint [(lbs/min)/(mg/L)]	Distance to Endpoint (miles)
0 - 4.4	0.1
4.4 - 37	0.2
37 - 97	0.3
97 - 180	0.4
180 - 340	0.6
340 - 530	0.8
530 - 760	1.0
760 - 1,000	1.2
1,000 - 1,500	1.4
1,500 - 1,900	1.6
1,900 - 2,400	1.8
2,400 - 2,900	2.0
2,900 - 3,500	2.2
3,500 - 4,400	2.4
4,400 - 5,100	2.6
5,100 - 5,900	2.8
5,900 - 6,800	3.0
6,800 - 7,700	3.2
7,700 - 9,000	3.4
9,000 - 10,000	3.6
10,000 - 11,000	3.8
11,000 - 12,000	4.0
12,000 - 14,000	4.2
14,000 - 15,000	4.4
15,000 - 16,000	4.6

Release Rate/Endpoint [(lbs/min)/(mg/L)]	Distance to Endpoint (miles)
16,000 - 18,000	4.8
18,000 - 19,000	5.0
19,000 - 21,000	5.2
21,000 - 23,000	5.4
23,000 - 24,000	5.6
24,000 - 26,000	5.8
26,000 - 28,000	6.0
28,000 - 29,600	6.2
29,600 - 35,600	6.8
35,600 - 42,000	7.5
42,000 - 48,800	8.1
48,800 - 56,000	8.7
56,000 - 63,600	9.3
63,600 - 71,500	9.9
71,500 - 88,500	11
88,500 - 107,000	12
107,000 - 126,000	14
126,000 - 147,000	15
147,000 - 169,000	16
169,000 - 191,000	17
191,000 - 215,000	19
215,000 - 279,000	22
279,000 - 347,000	25
>347,000	>25*

*Report distance as 25 miles

Reference Table 2
Neutrally Buoyant Plume Distances to Toxic Endpoint for Release Rate Divided by Endpoint
60-Minute Release, Rural Conditions, F Stability, Wind Speed 1.5 Meters per Second

Release Rate/Endpoint [(lbs/min)/(mg/L)]	Distance to Endpoint (miles)
0 - 5.5	0.1
5.5 - 46	0.2
46 - 120	0.3
120 - 220	0.4
220 - 420	0.6
420 - 650	0.8
650 - 910	1.0
910 - 1,200	1.2
1,200 - 1,600	1.4
1,600 - 1,900	1.6
1,900 - 2,300	1.8
2,300 - 2,600	2.0
2,600 - 2,900	2.2
2,900 - 3,400	2.4
3,400 - 3,700	2.6
3,700 - 4,100	2.8
4,100 - 4,400	3.0
4,400 - 4,800	3.2
4,800 - 5,200	3.4
5,200 - 5,600	3.6
5,600 - 5,900	3.8
5,900 - 6,200	4.0
6,200 - 6,700	4.2
6,700 - 7,000	4.4
7,000 - 7,400	4.6

Release Rate/Endpoint [(lbs/min)/(mg/L)]	Distance to Endpoint (miles)
7,400 - 7,700	4.8
7,700 - 8,100	5.0
8,100 - 8,500	5.2
8,500 - 8,900	5.4
8,900 - 9,200	5.6
9,200 - 9,600	5.8
9,600 - 10,000	6.0
10,000 - 10,400	6.2
10,400 - 11,700	6.8
11,700 - 13,100	7.5
13,100 - 14,500	8.1
14,500 - 15,900	8.7
15,900 - 17,500	9.3
17,500 - 19,100	9.9
19,100 - 22,600	11
22,600 - 26,300	12
26,300 - 30,300	14
30,300 - 34,500	15
34,500 - 38,900	16
38,900 - 43,600	17
43,600 - 48,400	19
48,400 - 61,500	22
61,500 - 75,600	25
>75,600	>25*

*Report distance as 25 miles

Reference Table 3
Neutrally Buoyant Plume Distances to Toxic Endpoint for Release Rate Divided by Endpoint 10-minute Release, Urban Conditions, F Stability, Wind Speed 1.5 Meters per Second

Release Rate/Endpoint [(lbs/min)/(mg/L)]	Distance to Endpoint (miles)
0 - 21	0.1
21 - 170	0.2
170 - 420	0.3
420 - 760	0.4
760 - 1,400	0.6
1,400 - 2,100	0.8
2,100 - 3,100	1.0
3,100 - 4,200	1.2
4,200 - 6,100	1.4
6,100 - 7,800	1.6
7,800 - 9,700	1.8
9,700 - 12,000	2.0
12,000 - 14,000	2.2
14,000 - 18,000	2.4
18,000 - 22,000	2.6
22,000 - 25,000	2.8
25,000 - 29,000	3.0
29,000 - 33,000	3.2
33,000 - 39,000	3.4
39,000 - 44,000	3.6
44,000 - 49,000	3.8
49,000 - 55,000	4.0
55,000 - 63,000	4.2
63,000 - 69,000	4.4
69,000 - 76,000	4.6

Release Rate/Endpoint [(lbs/min)/(mg/L)]	Distance to Endpoint (miles)
76,000 - 83,000	4.8
83,000 - 90,000	5.0
90,000 - 100,000	5.2
100,000 - 110,000	5.4
110,000 - 120,000	5.6
120,000 - 130,000	5.8
130,000 - 140,000	6.0
140,000 - 148,000	6.2
148,000 - 183,000	6.8
183,000 - 221,000	7.5
221,000 - 264,000	8.1
264,000 - 310,000	8.7
310,000 - 361,000	9.3
361,000 - 415,000	9.9
415,000 - 535,000	11
535,000 - 671,000	12
671,000 - 822,000	14
822,000 - 990,000	15
990,000 - 1,170,000	16
1,170,000 - 1,370,000	17
1,370,000 - 1,590,000	19
1,590,000 - 2,190,000	22
2,190,000 - 2,890,000	25
>2,890,000	>25*

*Report distance as 25 miles

Reference Table 4
Neutrally Buoyant Plume Distances to Toxic Endpoint for Release Rate Divided by Endpoint
60-Minute Release, Urban Conditions, F Stability, Wind Speed 1.5 Meters per Second

Release Rate/Endpoint [(lbs/min)/(mg/L)]	Distance to Endpoint (miles)
0 - 26	0.1
26 - 210	0.2
210 - 530	0.3
530 - 940	0.4
940 - 1,700	0.6
1,700 - 2,600	0.8
2,600 - 3,700	1.0
3,700 - 4,800	1.2
4,800 - 6,400	1.4
6,400 - 7,700	1.6
7,700 - 9,100	1.8
9,100 - 11,000	2.0
11,000 - 12,000	2.2
12,000 - 14,000	2.4
14,000 - 16,000	2.6
16,000 - 17,000	2.8
17,000 - 19,000	3.0
19,000 - 21,000	3.2
21,000 - 23,000	3.4
23,000 - 24,000	3.6
24,000 - 26,000	3.8
26,000 - 28,000	4.0
28,000 - 30,000	4.2
30,000 - 32,000	4.4
32,000 - 34,000	4.6
34,000 - 36,000	4.8
36,000 - 38,000	5.0
38,000 - 41,000	5.2
41,000 - 43,000	5.4
43,000 - 45,000	5.6
45,000 - 47,000	5.8
47,000 - 50,000	6.0
50,000 - 52,200	6.2
52,200 - 60,200	6.8
60,200 - 68,900	7.5
68,900 - 78,300	8.1
78,300 - 88,400	8.7
88,400 - 99,300	9.3
99,300 - 111,000	9.9
111,000 - 137,000	11
137,000 - 165,000	12
165,000 - 197,000	14
197,000 - 232,000	15
232,000 - 271,000	16
271,000 - 312,000	17
312,000 - 357,000	19
357,000 - 483,000	22
483,000 - 629,000	25
>629,000	>25*

*Report distance as 25 miles

Reference Table 5
Dense Gas Distances to Toxic Endpoint
10-minute Release, Rural Conditions, F Stability, Wind Speed 1.5 Meters per Second

Release Rate (lbs/min)	Toxic Endpoint (mg/L)															
	0.0004	0.0007	0.001	0.002	0.0035	0.005	0.0075	0.01	0.02	0.035	0.05	0.075	0.1	0.25	0.5	0.75
	Distance (Miles)															
1	2.2	1.7	1.5	1.1	0.8	0.7	0.5	0.5	0.3	0.2	0.2	0.2	0.1	0.1	#	#
2	3.0	2.4	2.1	1.5	1.1	0.9	0.7	0.7	0.4	0.3	0.3	0.2	0.2	0.1	<0.1	<0.1
5	4.8	3.7	3.0	2.2	1.7	1.5	1.2	1.0	0.7	0.5	0.4	0.3	0.3	0.2	0.1	0.1
10	6.8	5.0	4.2	3.0	2.4	2.1	1.7	1.4	1.0	0.7	0.6	0.5	0.4	0.2	0.2	0.1
30	11	8.7	6.8	5.2	3.9	3.4	2.8	2.4	1.7	1.3	1.1	0.9	0.7	0.4	0.3	0.2
50	14	11	9.3	6.8	5.0	4.2	3.5	3.0	2.2	1.7	1.4	1.1	0.9	0.6	0.4	0.3
100	19	15	12	8.7	6.8	5.8	4.8	4.2	2.9	2.2	1.9	1.6	1.3	0.8	0.5	0.4
150	24	18	15	11	8.1	6.8	5.7	5.0	3.6	2.7	2.3	1.9	1.6	0.9	0.6	0.5
250	>25	22	19	14	11	8.7	7.4	6.2	4.5	3.4	2.8	2.3	2.0	1.2	0.8	0.6
500	*	>25	>25	19	14	12	9.9	8.7	6.2	4.7	3.8	3.1	2.7	1.6	1.1	0.9
750	*	*	*	23	17	15	12	11	7.4	5.5	4.5	3.7	3.2	1.9	1.3	1.0
1,000	*	*	*	>25	20	17	14	12	8.1	6.2	5.2	4.2	3.6	2.2	1.4	1.1
1,500	*	*	*	*	24	20	16	14	9.9	7.4	6.2	5.0	4.3	2.5	1.7	1.3
2,000	*	*	*	*	>25	23	19	16	11	8.7	6.8	5.6	4.8	2.9	1.9	1.5
2,500	*	*	*	*	*	>25	20	18	12	9.3	8.1	6.2	5.3	3.2	2.1	1.6
3,000	*	*	*	*	*	*	23	20	14	9.9	8.7	6.8	5.6	3.4	2.2	1.7
4,000	*	*	*	*	*	*	>25	22	16	11	9.3	7.4	6.2	3.8	2.5	2.0
5,000	*	*	*	*	*	*	*	25	17	13	11	8.7	6.8	4.2	2.7	2.1
7,500	*	*	*	*	*	*	*	>25	20	15	12	9.9	8.7	4.9	3.2	2.5
10,000	*	*	*	*	*	*	*	*	24	17	14	11	9.3	5.5	3.6	2.8
15,000	*	*	*	*	*	*	*	*	>25	20	17	13	11	6.2	4.2	3.2
20,000	*	*	*	*	*	*	*	*	*	23	19	15	12	7.4	4.7	3.7
50,000	*	*	*	*	*	*	*	*	*	>25	>25	21	18	10	6.6	5.0
75,000	*	*	*	*	*	*	*	*	*	*	*	>25	21	12	7.6	5.8
100,000	*	*	*	*	*	*	*	*	*	*	*	*	24	13	8.5	6.4
150,000	*	*	*	*	*	*	*	*	*	*	*	*	>25	15	9.8	7.4
200,000	*	*	*	*	*	*	*	*	*	*	*	*	*	17	11	8.2

* >25 miles (report distance as 25 miles)

<0.1 mile (report distance as 0.1 mile)

Reference Table 6
Dense Gas Distances to Toxic Endpoint
60-minute Release, Rural Conditions, F Stability, Wind Speed 1.5 Meters per Second

Release Rate (lbs/min)	Toxic Endpoint (mg/L)															
	0.0004	0.0007	0.001	0.002	0.0035	0.005	0.0075	0.01	0.02	0.035	0.05	0.075	0.1	0.25	0.5	0.75
	Distance (Miles)															
1	3.7	2.7	2.2	1.4	1.0	0.8	0.6	0.5	0.3	0.2	0.2	0.1	0.1	<0.1	#	#
2	5.3	4.0	3.2	2.2	1.6	1.2	1.0	0.8	0.5	0.4	0.3	0.2	0.2	0.1	<0.1	<0.1
5	8.7	6.8	5.3	3.7	2.7	2.2	1.7	1.4	0.9	0.6	0.5	0.4	0.3	0.2	0.1	0.1
10	12	9.3	8.1	5.3	4.0	3.3	2.7	2.2	1.4	1.0	0.8	0.6	0.5	0.3	0.2	0.1
30	22	16	14	9.9	7.4	6.1	4.9	4.1	2.9	2.1	1.6	1.2	1.0	0.5	0.3	0.2
50	>25	21	18	12	9.3	8.1	6.2	5.4	3.8	2.7	2.2	1.7	1.4	0.7	0.4	0.3
100	*	>25	>25	18	13	11	9.3	7.4	5.5	4.0	3.2	2.5	2.1	1.1	0.7	0.5
150	*	*	*	22	17	14	11	9.9	6.8	4.9	4.0	3.1	2.7	1.4	0.9	0.6
250	*	*	*	>25	22	18	14	12	8.7	6.2	5.2	4.1	3.5	1.9	1.2	0.9
500	*	*	*	*	>25	25	20	17	12	9.3	7.4	5.8	5.0	2.9	1.8	1.3
750	*	*	*	*	*	>25	25	22	15	11	9.3	7.4	6.1	3.5	2.2	1.7
1,000	*	*	*	*	*	*	>25	25	17	12	11	8.1	6.8	4.0	2.6	2.0
1,500	*	*	*	*	*	*	*	>25	20	16	12	9.9	8.7	5.0	3.2	2.5
2,000	*	*	*	*	*	*	*	*	24	17	14	11	9.9	5.7	3.7	2.9
2,500	*	*	*	*	*	*	*	*	>25	20	16	13	11	6.2	4.2	3.2
3,000	*	*	*	*	*	*	*	*	*	21	17	14	12	6.8	4.5	3.5
4,000	*	*	*	*	*	*	*	*	*	24	20	16	14	8.1	5.2	4.0
5,000	*	*	*	*	*	*	*	*	*	>25	22	17	15	8.7	5.7	4.4
7,500	*	*	*	*	*	*	*	*	*	*	>25	21	18	11	6.8	5.2
10,000	*	*	*	*	*	*	*	*	*	*	*	24	20	12	7.4	6.0
15,000	*	*	*	*	*	*	*	*	*	*	*	>25	24	14	9.3	6.8
20,000	*	*	*	*	*	*	*	*	*	*	*	*	>25	16	9.9	8.1
50,000	*	*	*	*	*	*	*	*	*	*	*	*	*	22	14	11
75,000	*	*	*	*	*	*	*	*	*	*	*	*	*	>25	17	13
100,000	*	*	*	*	*	*	*	*	*	*	*	*	*	*	18	14
150,000	*	*	*	*	*	*	*	*	*	*	*	*	*	*	21	16
200,000	*	*	*	*	*	*	*	*	*	*	*	*	*	*	23	18

* >25 miles (report distance as 25 miles)

\# <0.1 mile (report distance as 0.1 mile)

Reference Table 7
Dense Gas Distances to Toxic Endpoint
10-minute Release, Urban Conditions, F Stability, Wind Speed 1.5 Meters per Second

Release Rate (lbs/min)	Toxic Endpoint (mg/L)															
	0.0004	0.0007	0.001	0.002	0.0035	0.005	0.0075	0.01	0.02	0.035	0.05	0.075	0.1	0.25	0.5	0.75
	Distance (Miles)															
1	1.6	1.2	1.1	0.7	0.6	0.4	0.4	0.3	0.2	0.2	0.1	0.1	0.1	#	#	#
2	2.2	1.7	1.4	1.1	0.8	0.6	0.5	0.4	0.3	0.2	0.2	0.1	0.1	<0.1	#	#
5	3.5	2.7	2.2	1.6	1.2	1.0	0.8	0.7	0.5	0.4	0.3	0.2	0.2	0.1	<0.1	#
10	4.9	3.8	3.1	2.2	1.7	1.4	1.2	1.0	0.7	0.5	0.4	0.3	0.2	0.1	0.1	<0.1
30	8.1	6.2	5.3	3.7	2.9	2.4	2.0	1.7	1.2	0.9	0.7	0.6	0.4	0.2	0.1	0.1
50	11	8.1	6.8	4.8	3.7	3.1	2.5	2.1	1.5	1.1	0.9	0.7	0.6	0.3	0.2	0.1
100	15	11	9.3	6.8	5.2	4.2	3.5	3.0	2.1	1.6	1.3	1.0	0.9	0.5	0.3	0.2
150	19	14	12	8.1	6.1	5.2	4.3	3.6	2.5	1.9	1.6	1.2	1.1	0.6	0.4	0.2
250	24	18	15	11	8.1	6.8	5.4	4.6	3.3	2.4	2.0	1.6	1.4	0.7	0.5	0.3
500	>25	>25	21	15	11	9.3	7.4	6.2	4.5	3.4	2.8	2.2	1.9	1.1	0.7	0.5
750	*	*	>25	18	14	11	9.3	8.1	5.5	4.1	3.3	2.6	2.2	1.3	0.8	0.6
1,000	*	*	*	21	16	13	11	9.3	6.2	4.6	3.8	3.0	2.5	1.5	0.9	0.7
1,500	*	*	*	>25	19	16	12	11	7.4	5.6	4.6	3.7	3.0	1.7	1.1	0.8
2,000	*	*	*	*	22	18	15	12	8.7	6.2	5.2	4.1	3.5	2.0	1.3	0.9
2,500	*	*	*	*	24	20	16	14	9.9	6.8	5.8	4.7	3.8	2.2	1.4	1.1
3,000	*	*	*	*	>25	22	18	16	11	7.4	6.2	5.0	4.2	2.4	1.6	1.2
4,000	*	*	*	*	*	25	20	17	12	8.7	6.8	5.6	4.8	2.7	1.7	1.3
5,000	*	*	*	*	*	>25	23	20	14	9.9	8.1	6.2	5.3	3.0	1.9	1.4
7,500	*	*	*	*	*	*	>25	24	16	12	9.9	7.4	6.2	3.6	2.3	1.7
10,000	*	*	*	*	*	*	*	>25	19	14	11	8.7	7.4	4.1	2.6	2.0
15,000	*	*	*	*	*	*	*	*	22	16	13	11	8.7	4.9	3.1	2.3
20,000	*	*	*	*	*	*	*	*	>25	19	15	12	9.9	5.5	3.5	2.7
50,000	*	*	*	*	*	*	*	*	*	>25	23	17	15	8.1	5.1	3.8
75,000	*	*	*	*	*	*	*	*	*	*	>25	21	17	9.6	6.0	4.5
100,000	*	*	*	*	*	*	*	*	*	*	*	24	20	11	6.8	5.1
150,000	*	*	*	*	*	*	*	*	*	*	*	>25	23	13	8.1	6.1
200,000	*	*	*	*	*	*	*	*	*	*	*	*	>25	14	8.9	6.7

* >25 miles (report distance as 25 miles)

\# <0.1 mile (report distance as 0.1 mile)

Reference Table 8
Dense Gas Distances to Toxic Endpoint
60-minute Release, Urban Conditions, F Stability, Wind Speed 1.5 Meters per Second

Release Rate (lbs/min)	Toxic Endpoint (mg/L)															
	0.0004	0.0007	0.001	0.002	0.0035	0.005	0.0075	0.01	0.02	0.035	0.05	0.075	0.1	0.25	0.5	0.75
	Distance (Miles)															
1	2.6	1.9	1.5	1.1	0.7	0.6	0.4	0.4	0.2	0.2	0.1	0.1	0.1	#	#	#
2	3.8	2.9	2.3	1.5	1.1	0.9	0.7	0.6	0.4	0.2	0.2	0.1	0.1	<0.1	#	#
5	6.2	4.7	3.9	2.6	1.9	1.5	1.2	0.9	0.6	0.4	0.3	0.2	0.2	0.1	<0.1	#
10	9.3	6.8	5.6	3.9	2.9	2.3	1.8	1.5	0.9	0.7	0.5	0.4	0.3	0.2	0.1	<0.1
30	16	12	9.9	7.4	5.3	4.3	3.4	2.9	1.9	1.3	1.0	0.7	0.6	0.3	0.2	0.1
50	22	16	14	9.3	6.8	5.7	4.5	3.8	2.6	1.8	1.4	1.1	0.9	0.4	0.2	0.2
100	>25	24	20	14	9.9	8.1	6.8	5.7	3.8	2.7	2.2	1.7	1.4	0.7	0.4	0.3
150	*	>25	24	17	12	11	8.1	6.8	4.8	3.5	2.8	2.2	1.8	0.9	0.5	0.3
250	*	*	>25	22	16	14	11	9.3	6.2	4.5	3.7	2.9	2.4	1.2	0.7	0.5
500	*	*	*	>25	24	19	16	13	9.3	6.8	5.4	4.2	3.5	1.9	1.1	0.7
750	*	*	*	*	>25	24	19	16	11	8.1	6.8	5.2	4.3	2.4	1.4	1.0
1,000	*	*	*	*	*	>25	22	19	13	9.3	7.4	6.0	5.0	2.8	1.6	1.2
1,500	*	*	*	*	*	*	>25	24	16	12	9.3	7.4	6.2	3.4	2.1	1.5
2,000	*	*	*	*	*	*	*	>25	19	13	11	8.7	7.4	4.0	2.5	1.8
2,500	*	*	*	*	*	*	*	*	20	15	12	9.3	8.1	4.5	2.8	2.1
3,000	*	*	*	*	*	*	*	*	22	16	13	11	8.7	4.9	3.0	2.2
4,000	*	*	*	*	*	*	*	*	>25	19	16	12	9.9	5.6	3.5	2.6
5,000	*	*	*	*	*	*	*	*	*	21	17	14	11	6.2	4.0	3.0
7,500	*	*	*	*	*	*	*	*	*	>25	20	16	14	7.4	4.8	3.6
10,000	*	*	*	*	*	*	*	*	*	*	24	19	16	8.7	5.5	4.2
15,000	*	*	*	*	*	*	*	*	*	*	>25	22	19	11	6.8	5.1
20,000	*	*	*	*	*	*	*	*	*	*	*	>25	21	12	7.4	5.8
50,000	*	*	*	*	*	*	*	*	*	*	*	*	>25	18	11	8.7
75,000	*	*	*	*	*	*	*	*	*	*	*	*	*	21	13	10
100,000	*	*	*	*	*	*	*	*	*	*	*	*	*	24	15	11
150,000	*	*	*	*	*	*	*	*	*	*	*	*	*	>25	18	14
200,000	*	*	*	*	*	*	*	*	*	*	*	*	*	*	20	15

* > 25 miles (report distance as 25 miles)

\# <0.1 mile (report distance as 0.1 mile)

Reference Table 9
Distances to Toxic Endpoint for Anhydrous Ammonia Liquefied Under Pressure
F Stability, Wind Speed 1.5 Meters per Second

Release Rate (lbs/min)	Distance to Endpoint (miles)	
	Rural	Urban
1	0.1	<0.1*
2	0.1	0.1
5	0.1	0.1
10	0.2	0.1
15	0.2	0.2
20	0.3	0.2
30	0.3	0.2
40	0.4	0.3
50	0.4	0.3
60	0.5	0.3
70	0.5	0.3
80	0.5	0.4
90	0.6	0.4
100	0.6	0.4
150	0.7	0.5
200	0.8	0.6
250	0.9	0.6
300	1.0	0.7
400	1.2	0.8
500	1.3	0.9
600	1.4	0.9
700	1.5	1.0
750	1.6	1.0
800	1.6	1.1
900	1.7	1.2

*Report distance as 0.1 mile

Release Rate (lbs/min)	Distance to Endpoint (miles)	
	Rural	Urban
1,000	1.8	1.2
1,500	2.2	1.5
2,000	2.6	1.7
2,500	2.9	1.9
3,000	3.1	2.0
4,000	3.6	2.3
5,000	4.0	2.6
6,000	4.4	2.8
7,000	4.7	3.1
7,500	4.9	3.2
8,000	5.1	3.3
9,000	5.4	3.4
10,000	5.6	3.6
15,000	6.9	4.4
20,000	8.0	5.0
25,000	8.9	5.6
30,000	9.7	6.1
40,000	11	7.0
50,000	12	7.8
75,000	15	9.5
100,000	18	10
150,000	22	13
200,000	**	15
250,000	**	17
750,000	**	**

** More than 25 miles (report distance as 25 miles)

Reference Table 10
Distances to Toxic Endpoint for Non-liquefied Ammonia, Ammonia Liquefied by Refrigeration, or Aqueous Ammonia
F Stability, Wind Speed 1.5 Meters per Second

Release Rate (lbs/min)	Distance to Endpoint (miles)	
	Rural	Urban
1	0.1	<0.1*
2	0.1	
5	0.1	
10	0.2	0.1
15	0.2	0.1
20	0.3	0.1
30	0.3	0.1
40	0.4	0.1
50	0.4	0.1
60	0.4	0.2
70	0.5	0.2
80	0.5	0.2
90	0.5	0.2
100	0.6	0.2
150	0.7	0.2
200	0.8	0.3
250	0.8	0.3
300	0.9	0.3
400	1.1	0.4
500	1.2	0.4
600	1.3	0.4
700	1.4	0.5
750	1.4	0.5
800	1.5	0.5
900	1.5	0.6
1,000	1.6	0.6
1,500	2.0	0.7
2,000	2.2	0.8
2,500	2.5	0.9
3,000	2.7	1.0
4,000	3.1	1.1
5,000	3.4	1.2
6,000	3.7	1.3
7,000	4.0	1.4
7,500	4.1	1.5
8,000	4.2	1.5
9,000	4.5	1.6
10,000	4.7	1.7
15,000	5.6	2.0
20,000	6.5	2.4
25,000	7.2	2.6
30,000	7.8	2.8
40,000	8.9	3.3
50,000	9.8	3.6
75,000	12	4.4
100,000	14	5.0
150,000	16	6.1
200,000	19	7.0
250,000	21	7.8
750,000	**	13

*Report distance as 0.1 mile

** More than 25 miles (report distance as 25 miles)

Reference Table 11
Distances to Toxic Endpoint for Chlorine
F Stability, Wind Speed 1.5 Meters per Second

Release Rate (lbs/min)	Distance to Endpoint (miles)	
	Rural	Urban
1	0.2	0.1
2	0.3	0.1
5	0.5	0.2
10	0.7	0.3
15	0.8	0.4
20	1.0	0.4
30	1.2	0.5
40	1.4	0.6
50	1.5	0.6
60	1.7	0.7
70	1.8	0.8
80	1.9	0.8
90	2.0	0.9
100	2.2	0.9
150	2.6	1.2
200	3.0	1.3
250	3.4	1.5
300	3.7	1.6
400	4.2	1.9
500	4.7	2.1
600	5.2	2.3
700	5.6	2.5

Release Rate (lbs/min)	Distance to Endpoint (miles)	
	Rural	Urban
750	5.8	2.6
800	5.9	2.7
900	6.3	2.9
1,000	6.6	3.0
1,500	8.1	3.8
2,000	9.3	4.4
2,500	10	4.9
3,000	11	5.4
4,000	13	6.2
5,000	14	7.0
6,000	16	7.6
7,000	17	8.3
7,500	18	8.6
8,000	18	8.9
9,000	19	9.4
10,000	20	9.9
15,000	25	12
20,000	*	14
25,000	*	16
30,000	*	18
40,000	*	20
50,000	*	*

* More than 25 miles (report distance as 25 miles)

Reference Table 12
Distances to Toxic Endpoint for Anhydrous Sulfur Dioxide
F Stability, Wind Speed 1.5 Meters per Second

Release Rate (lbs/min)	Distance to Endpoint (miles)	
	Rural	Urban
1	0.2	0.1
2	0.2	0.1
5	0.4	0.2
10	0.6	0.2
15	0.7	0.3
20	0.9	0.4
30	1.1	0.5
40	1.3	0.5
50	1.4	0.6
60	1.6	0.7
70	1.8	0.7
80	1.9	0.8
90	2.0	0.8
100	2.1	0.9
150	2.7	1.1
200	3.1	1.3
250	3.6	1.4
300	3.9	1.6
400	4.6	1.9
500	5.2	2.1
600	5.8	2.3
700	6.3	2.5

Release Rate (lbs/min)	Distance to Endpoint (miles)	
	Rural	Urban
750	6.6	2.6
800	6.8	2.7
900	7.2	2.9
1,000	7.7	3.1
1,500	9.6	3.8
2,000	11	4.5
2,500	13	5.0
3,000	14	5.6
4,000	17	6.5
5,000	19	7.3
6,000	21	8.1
7,000	23	8.8
7,500	24	9.1
8,000	25	9.5
9,000	*	10
10,000	*	11
15,000	*	13
20,000	*	16
25,000	*	18
30,000	*	19
40,000	*	23
50,000	*	*

* More than 25 miles (report distance as 25 miles)

Reference Table 13
Distance to Overpressure of 1.0 psi for Vapor Cloud Explosions of 500 - 2,000,000 Pounds of Regulated Flammable Substances Based on TNT Equivalent Method, 10 Percent Yield Factor

	Quantity in Cloud (pounds)	500	2,000	5,000	10,000	20,000	50,000	100,000	200,000	500,000	1,000,000	2,000,000
CAS No.	Chemical Name	Distance (Miles) to 1 psi Overpressure										
75-07-0	Acetaldehyde	0.05	0.08	0.1	0.1	0.2	0.2	0.3	0.4	0.5	0.7	0.8
74-86-2	Acetylene	0.07	0.1	0.1	0.2	0.2	0.3	0.4	0.5	0.7	0.8	1.0
598-73-2	Bromotrifluoroethylene	0.02	0.04	0.05	0.06	0.08	0.1	0.1	0.2	0.2	0.3	0.4
106-99-0	1,3-Butadiene	0.06	0.1	0.1	0.2	0.2	0.3	0.4	0.5	0.6	0.8	1.0
106-97-8	Butane	0.06	0.1	0.1	0.2	0.2	0.3	0.4	0.5	0.6	0.8	1.0
25167-67-3	Butene	0.06	0.1	0.1	0.2	0.2	0.3	0.4	0.5	0.6	0.8	1.0
590-18-1	2-Butene-cis	0.06	0.1	0.1	0.2	0.2	0.3	0.4	0.5	0.6	0.8	1.0
624-64-6	2-Butene-trans	0.06	0.1	0.1	0.2	0.2	0.3	0.4	0.5	0.6	0.8	1.0
106-98-9	1-Butene	0.06	0.1	0.1	0.2	0.2	0.3	0.4	0.5	0.6	0.8	1.0
107-01-7	2-Butene	0.06	0.1	0.1	0.2	0.2	0.3	0.4	0.5	0.6	0.8	1.0
463-58-1	Carbon oxysulfide	0.04	0.06	0.08	0.1	0.1	0.2	0.2	0.3	0.4	0.5	0.6
7791-21-1	Chlorine monoxide	0.02	0.03	0.04	0.05	0.06	0.08	0.1	0.1	0.2	0.2	0.3
590-21-6	1-Chloropropylene	0.05	0.08	0.1	0.1	0.2	0.2	0.3	0.4	0.5	0.6	0.8
557-98-2	2-Chloropropylene	0.05	0.08	0.1	0.1	0.2	0.2	0.3	0.4	0.5	0.6	0.8
460-19-5	Cyanogen	0.05	0.08	0.1	0.1	0.2	0.2	0.3	0.4	0.5	0.6	0.8
75-19-4	Cyclopropane	0.06	0.1	0.1	0.2	0.2	0.3	0.4	0.5	0.6	0.8	1.0
4109-96-0	Dichlorosilane	0.04	0.06	0.08	0.1	0.1	0.2	0.2	0.3	0.4	0.5	0.6
75-37-6	Difluoroethane	0.04	0.06	0.09	0.1	0.1	0.2	0.2	0.3	0.4	0.5	0.6
124-40-3	Dimethylamine	0.06	0.09	0.1	0.2	0.2	0.3	0.3	0.4	0.6	0.7	0.9
463-82-1	2,2-Dimethylpropane	0.06	0.1	0.1	0.2	0.2	0.3	0.4	0.5	0.6	0.8	1.0
74-84-0	Ethane	0.06	0.1	0.1	0.2	0.2	0.3	0.4	0.5	0.6	0.8	1.0
107-00-6	Ethyl acetylene	0.06	0.1	0.1	0.2	0.2	0.3	0.4	0.5	0.6	0.8	1.0
75-04-7	Ethylamine	0.06	0.09	0.1	0.2	0.2	0.3	0.3	0.4	0.6	0.7	0.9

Reference Table 13 (continued)

CAS No.	Quantity in Cloud (pounds) / Chemical Name	500	2,000	5,000	10,000	20,000	50,000	100,000	200,000	500,000	1,000,000	2,000,000
		Distance (Miles) to 1 psi Overpressure										
75-00-3	Ethyl chloride	0.05	0.08	0.1	0.1	0.2	0.2	0.3	0.4	0.5	0.6	0.8
74-85-1	Ethylene	0.06	0.1	0.1	0.2	0.2	0.3	0.4	0.5	0.7	0.8	1.0
60-29-7	Ethyl ether	0.06	0.09	0.1	0.2	0.2	0.3	0.3	0.4	0.6	0.7	0.9
75-08-1	Ethyl mercaptan	0.05	0.09	0.1	0.2	0.2	0.2	0.3	0.4	0.5	0.7	0.9
109-95-5	Ethyl nitrite	0.05	0.07	0.1	0.1	0.2	0.2	0.3	0.3	0.5	0.6	0.7
1333-74-0	Hydrogen	0.09	0.1	0.2	0.2	0.3	0.4	0.5	0.6	0.9	1.1	1.4
75-28-5	Isobutane	0.06	0.1	0.1	0.2	0.2	0.3	0.4	0.5	0.6	0.8	1.0
78-78-4	Isopentane	0.06	0.1	0.1	0.2	0.2	0.3	0.4	0.5	0.6	0.8	1.0
78-79-5	Isoprene	0.06	0.1	0.1	0.2	0.2	0.3	0.4	0.5	0.6	0.8	1.0
75-31-0	Isopropylamine	0.06	0.09	0.1	0.2	0.2	0.3	0.3	0.4	0.6	0.7	0.9
75-29-6	Isopropyl chloride	0.05	0.08	0.1	0.1	0.2	0.2	0.3	0.4	0.5	0.6	0.8
74-82-8	Methane	0.07	0.1	0.1	0.2	0.2	0.3	0.4	0.5	0.7	0.8	1.0
74-89-5	Methylamine	0.06	0.09	0.1	0.2	0.2	0.3	0.3	0.4	0.6	0.7	0.9
563-45-1	3-Methyl-1-butene	0.06	0.1	0.1	0.2	0.2	0.3	0.4	0.5	0.6	0.8	1.0
563-46-2	2-Methyl-1-butene	0.06	0.1	0.1	0.2	0.2	0.3	0.4	0.5	0.6	0.8	1.0
115-10-6	Methyl ether	0.05	0.09	0.1	0.1	0.2	0.3	0.3	0.4	0.5	0.7	0.9
107-31-3	Methyl formate	0.04	0.07	0.1	0.1	0.2	0.2	0.3	0.3	0.4	0.6	0.7
115-11-7	2-Methylpropene	0.06	0.1	0.1	0.2	0.2	0.3	0.4	0.5	0.6	0.8	1.0
504-60-9	1,3-Pentadiene	0.06	0.1	0.1	0.2	0.2	0.3	0.4	0.5	0.6	0.8	1.0
109-66-0	Pentane	0.06	0.1	0.1	0.2	0.2	0.3	0.4	0.5	0.6	0.8	1.0
109-67-1	1-Pentene	0.06	0.1	0.1	0.2	0.2	0.3	0.4	0.5	0.6	0.8	1.0
646-04-8	2-Pentene, (E)-	0.06	0.1	0.1	0.2	0.2	0.3	0.4	0.5	0.6	0.8	1.0
627-20-3	2-Pentene, (Z)-	0.06	0.1	0.1	0.2	0.2	0.3	0.4	0.5	0.6	0.8	1.0
463-49-0	Propadiene	0.06	0.1	0.1	0.2	0.2	0.3	0.4	0.5	0.6	0.8	1.0

Reference Table 13 (continued)

Quantity in Cloud (pounds)		500	2,000	5,000	10,000	20,000	50,000	100,000	200,000	500,000	1,000,000	2,000,000
CAS No.	Chemical Name	Distance (Miles) to 1 psi Overpressure										
74-98-6	Propane	0.06	0.1	0.1	0.2	0.2	0.3	0.4	0.5	0.6	0.8	1.0
115-07-1	Propylene	0.06	0.1	0.1	0.2	0.2	0.3	0.4	0.5	0.6	0.8	1.0
74-99-7	Propyne	0.06	0.1	0.1	0.2	0.2	0.3	0.4	0.5	0.6	0.8	1.0
7803-62-5	Silane	0.06	0.1	0.1	0.2	0.2	0.3	0.4	0.5	0.6	0.8	1.0
116-14-3	Tetrafluoroethylene	0.02	0.03	0.04	0.05	0.07	0.09	0.1	0.1	0.2	0.2	0.3
75-76-3	Tetramethylsilane	0.06	0.1	0.1	0.2	0.2	0.3	0.4	0.5	0.6	0.8	1.0
10025-78-2	Trichlorosilane	0.03	0.04	0.06	0.08	0.1	0.1	0.2	0.2	0.3	0.4	0.4
79-38-9	Trifluorochloroethylene	0.02	0.03	0.05	0.06	0.07	0.1	0.1	0.2	0.2	0.3	0.3
75-50-3	Trimethylamine	0.06	0.1	0.1	0.2	0.2	0.3	0.4	0.4	0.6	0.8	1.0
689-97-4	Vinyl acetylene	0.06	0.1	0.1	0.2	0.2	0.3	0.4	0.5	0.6	0.8	1.0
75-01-4	Vinyl chloride	0.05	0.08	0.1	0.1	0.2	0.2	0.3	0.4	0.5	0.6	0.8
109-92-2	Vinyl ethyl ether	0.06	0.09	0.1	0.2	0.2	0.3	0.3	0.4	0.6	0.7	0.9
75-02-5	Vinyl fluoride	0.02	0.04	0.05	0.06	0.08	0.1	0.1	0.2	0.2	0.3	0.4
75-35-4	Vinylidene chloride	0.04	0.06	0.08	0.1	0.1	0.2	0.2	0.3	0.4	0.5	0.6
75-38-7	Vinylidene fluoride	0.04	0.06	0.09	0.1	0.1	0.2	0.2	0.3	0.4	0.5	0.6
107-25-5	Vinyl methyl ether	0.06	0.09	0.1	0.2	0.2	0.3	0.3	0.4	0.6	0.7	0.9

6 DETERMINING ALTERNATIVE RELEASE SCENARIOS

In Chapter 6

- Considerations for alternative release scenarios for regulated substances in Program 2 or Program 3 processes.
- Potential alternative scenarios for releases of flammable substances.

You are required to analyze at least one alternative release scenario for each listed toxic substance you have in a Program 2 or Program 3 process above its threshold quantity. You also are required to analyze one alternative release scenario for flammable substances in Program 2 or 3 processes as a class (i.e., you analyze one scenario involving a flammable substance as a representative scenario for all the regulated flammable substances you have on site in Program 2 or Program 3 processes). You do not need to analyze an alternative scenario for each flammable substance. For example, if you have five listed substances – chlorine, ammonia, hydrogen chloride, propane, and acetylene – above the threshold in Program 2 or 3 processes, you will need to analyze one alternative scenario each for chlorine, ammonia, and hydrogen chloride and a single alternative scenario to cover propane and acetylene (listed flammable substances). Even if you have a substance above the threshold in several processes or locations, you need only analyze one alternative scenario for it.

According to the rule (40 CFR 68.28), alternative scenarios should be more likely to occur than the worst-case scenario and should reach an endpoint offsite, unless no such scenario exists. Release scenarios considered should include, but are not limited to, the following:

- Transfer hose releases due to splits or sudden hose uncoupling;
- Process piping releases from failures at flanges, joints, welds, valves and valve seals, and drains or bleeds;
- Process vessel or pump releases due to cracks, seal failure, or drain, bleed, or plug failure;
- Vessel overfilling and spill, or overpressurization and venting through relief valves or rupture disks; and
- Shipping container mishandling or puncturing leading to a spill.

Alternative release scenarios for toxic substances should be those that lead to concentrations above the toxic endpoint beyond your fenceline. Scenarios for flammable substances should have the potential to cause substantial damage, including on-site damage. Those releases that have the potential to reach the public are of the greatest concern. You should consider unusual situations, such as start-up and shut-down, in selecting an appropriate alternative scenario.

For alternative release scenarios, you are allowed to consider active mitigation systems, such as interlocks, shutdown systems, pressure relieving devices, flares, emergency isolation systems, and fire water and deluge systems, as well as passive mitigation systems, as described in Sections 3.1.2 and 3.2.3.

For alternative release scenarios for regulated substances used in ammonia refrigeration, chemical distribution, propane distribution, warehouses, or POTWs, consult EPA's risk management program guidance documents for these industry sectors.

You have a number of options for selecting release scenarios for toxic or flammable substances.

- You may use your worst-case release scenario and apply your active mitigation system to limit the quantity released and the duration of the release.
- You may use information from your process hazards analysis, if you have conducted one, to select a scenario.
- You may review your accident history and choose an actual event as the basis of your scenario.
- If you have not conducted a process hazards analysis, you may review your operations and identify possible events and failures.

Whichever approach you select, the key information you need to define is the quantity to be released and the time over which it will be released; together, these allow you to estimate the release rate and use essentially the same methods you used for the worst-case analysis.

For flammable substances , the choice of alternative release scenarios is somewhat more complicated than for toxic substances, because the consequences of a release and the endpoint of concern may vary. For the flammable worst case, the consequence of concern is a vapor cloud explosion, with an overpressure endpoint. For alternative scenarios (e.g., fires), other endpoints (e.g., heat radiation) may need to be considered.

Possible scenarios involving flammable substances include:

- Vapor cloud fires (flash fires) may result from dispersion of a cloud of flammable vapor and ignition of the cloud following dispersion. Such a fire could flash back and could represent a severe heat radiation hazard to anyone in the area of the cloud. This guidance provides methods to estimate distances to a concentration equal to the lower flammability limit (LFL) for this type of fire. (See Sections 9.1, 9.2, and 10.1.)
- A pool fire, with potential radiant heat effects, may result from a spill of a flammable liquid. This guidance provides a simple method for estimating the distance from a pool fire to a radiant heat level that could cause second degree burns from a 40-second exposure. (See Section 10.2).
- A boiling liquid, expanding vapor explosion (BLEVE), leading to a fireball that may produce intense heat, may occur if a vessel containing flammable material ruptures explosively as a result of exposure to fire. Heat radiation from the fireball is the primary hazard; vessel fragments and overpressure from the explosion also can result. BLEVEs are generally considered unlikely events; however, if you think a BLEVE is possible at your site, this guidance provides a method to estimate the distance at which radiant heat effects might lead to second degree burns. (See Section 10.3.) You also may want to consider models or

calculation methods to estimate effects of vessel fragmentation. (See Appendix A for references that may provide useful information for estimating such effects.)

- For a vapor cloud explosion to occur, rapid release of a large quantity, turbulent conditions (caused by a turbulent release or congested conditions in the area of the release, or both), and other factors are generally necessary. Vapor cloud explosions generally are considered unlikely events; however, if conditions at your site are conducive to vapor cloud explosions, you may want to consider a vapor cloud explosion as an alternative scenario. This guidance provides methods you may use to estimate the distance to 1 psi overpressure for a vapor cloud detonation, based on less conservative assumptions than the worst-case analysis. (See Section 10.4.) A vapor cloud deflagration, involving lower flame speeds than a detonation and resulting in less damaging blast effects, is more likely than a detonation. This guidance does not provide methods for estimating the effects of a deflagration, but you may use other methods of analysis if you want to consider such events. (See Appendix A for references that may provide useful information.)

- A jet fire may result from the puncture or rupture of a tank or pipeline containing a compressed or liquefied gas under pressure. The gas discharging from the hole can form a jet that "blows" into the air in the direction of the hole; the jet then may ignite. Jet fires could contribute to BLEVEs and fireballs if they impinge on tanks of flammable substances. A large horizontal jet fire may have the potential to pose an offsite hazard. This guidance does not include a method for estimating consequence distances for jet fires. If you want to consider a jet fire as an alternative scenario, you should consider other models or methods for the consequence analysis. (See Appendix A for references that may provide useful information.)

If you carry out an alternative scenario analysis for a flammable mixture (i.e., a mixture that meets the criteria for NFPA 4), you need to consider all flammable components of the mixture, not just the regulated flammable substance or substances in the mixture (see Section 5.2 on flammable mixtures). If the mixture contains both flammable and non-flammable components, the analysis should be carried out considering only the flammable components.

Chapter 7 provides detailed information on calculating release rates for alternative release scenarios for toxic substances. If you can estimate release rates for the toxic gases and liquids you have on site based on readily available information, you may skip Chapter 7 and go to Chapter 8. Chapter 8 describes how to estimate distances to the toxic endpoint for alternative scenarios for toxic substances. Chapters 9 provides information on calculation release rates for flammable substances. Chapter 10 describes how to estimate distances to flammable endpoints.

This page intentionally left blank.

7 ESTIMATION OF RELEASE RATES FOR ALTERNATIVE SCENARIOS FOR TOXIC SUBSTANCES

For the alternative scenario analysis, you may use typical meteorological conditions and typical ambient temperature and humidity for your site. This guidance assumes D atmospheric stability and wind speed of 3.0 meters per second (6.7 miles per hour) as conditions likely to be applicable to many sites.

7.1 Release Rates for Toxic Gases

<u>In Section 7.1</u>

- 7.1.1 Methods for unmitigated releases of toxic gases, including:
  - -- Release of toxic gas from a hole in a tank or pipe (for choked flow conditions, or maximum flow rate),
  - -- Release of toxic gas from a pipe, based on the flow rate through the pipe, or based on a hole in the pipe (using the same method as for a hole in a tank),
  - -- Puff releases (no method is provided; users are directed to use other methods),
  - -- Gases liquefied under pressure, including gaseous releases from holes above the liquid level in the tank and releases from holes in the liquid space, and
  - -- Consideration of duration of releases of toxic gas.
- 7.1.2 Methods for adjusting the estimated release rate to account for active or passive mitigation, including:
  - -- Active mitigation to reduce the release duration (e.g., automatic shutoff valves),
  - -- Active mitigation to reduce the release rate to air, and
  - -- Passive mitigation (using the same method as for worst-case scenarios).

7.1.1 Unmitigated Releases of Toxic Gases

Gaseous Releases

<u>Gaseous Release from Tank.</u> Instead of assuming release of the entire contents of a vessel containing a toxic gas, you may decide to consider a more likely scenario as developed by the process hazards analysis, such as release from a hole in a vessel or pipe. To estimate a hole size you might assume, for example, the hole size that would result from shearing off a valve or pipe from a vessel containing a regulated substance. If you have a gas leak from a tank, you may use the following simplified equation to estimate a release rate based on hole size, tank pressure, and the properties of the gas. This equation applies to choked flow, or maximum gas flow rate. Choked flow generally would be expected for gases under pressure. (See Appendix D, Section D.6 for the derivation of this equation.)

$$QR = HA \times P_t \times \frac{1}{\sqrt{T_t}} \times GF \qquad (7\text{-}1)$$

where:

QR	=	Release rate (pounds per minute)
HA	=	Hole or puncture area (square inches) (from hazard evaluation or best estimate)
P_t	=	Tank pressure (pounds per square inch absolute (psia)) (from process information; for liquefied gases, equilibrium vapor pressure at 25 °C is included in Exhibit B-1, Appendix B)
T_t	=	Tank temperature (K), where K is absolute temperature in kelvins; 25 °C (77 °F) is 298 K
GF	=	Gas Factor, incorporating discharge coefficient, ratio of specific heats, molecular weight, and conversion factors (listed for each regulated toxic gas in Exhibit B-1, Appendix B)

You can estimate the hole area from the size and shape of the hole. For a circular hole, you would use the formula for the area of a circle (area = πr^2, where π is 3.14 and r is the radius of the circle; the radius is half the diameter).

This equation will give an estimate of the initial release rate. It will overestimate the overall release rate, because it does not take into account the decrease in the release rate as the pressure in the tank decreases. You may use a computer model or another calculation method if you want a more realistic estimate of the release rate. As discussed below, you may use this equation for releases of gases liquefied under pressure if the release would be primarily gas (e.g., if the hole is in the head space of the tank, well above the liquid level).

Example 19. Release of Toxic Gas from Tank (Diborane)

You have a tank that contains diborane gas at a pressure of 30 psia. The temperature of the tank and its contents is 298 K (25 °C). A valve on the side of the tank shears off, leaving a circular hole 2 ½ inches in diameter in the tank wall. You estimate the area from the formula for area of a circle (πr^2, where r is the radius). The radius of the hole is 1 1/4 inches, so the area is $\pi \times (1\ 1/4)^2$, or 5 square inches. From Exhibit B-1, the Gas Factor for diborane is 17. Therefore, the release rate, from Equation 7-1, is:

$$QR = 5 \times 30 \times 1/(298)^{½} \times 17 = 148 \text{ pounds per minute}$$

<u>Gaseous Release from Pipe</u>. If shearing of a pipe may be an alternative scenario for a toxic gas at your site, you could use the usual flow rate through the pipe as the release rate and carry out the estimation of distance as discussed in Chapter 8.

If you want to consider a release of toxic gas through a hole in a pipe as an alternative scenario, you may use the method described above for a gas release from a hole in a tank. This method neglects the effects of friction along the pipe and, therefore, provides a conservative estimate of the release rate.

Puff Releases. If a gaseous release from a hole in a tank or pipe is likely to be stopped very quickly (e.g., by a block valve), resulting in a puff of toxic gas that forms a vapor cloud rather than a plume, you may want to consider other methods for determining a consequence distance. A cloud of toxic gas resulting from a puff release will not exhibit the same behavior as a plume resulting from a longer release (e.g., a release over 10 minutes).

Liquefied Gases

Gases Liquefied Under Pressure. Gases liquefied under pressure may be released as gases, liquids, or a combination (two-phase), depending on a number of factors, including liquid level and the location of the hole relative to the liquid level. The resulting impact distances can vary greatly.

For releases from holes above the liquid level in a tank of gas liquefied under pressure, the release could be primarily gas, or the release may involve rapid vaporization of a fraction of the liquefied gas and possibly aerosolization (two-phase release). It is complex to determine which type of release (i.e., gas, two-phase) will occur and the likely mix of gases and liquids in a two-phase release. The methods presented in this guidance do not definitively address this situation. As a rule of thumb, if the head space is large and the distance between the hole and the liquid level is relatively large given the height of the tank or vessel, you could assume the release is gaseous and, therefore, use Equation 7-1 above. (Exhibit B-1, Appendix B, includes the equilibrium vapor pressure in psia for listed toxic gases liquefied under pressure at 25 °C; this pressure can be used in Equation 7-1.) However, use of this equation will not be conservative if the head space is small and the release from the hole is two-phased. In situations where you are unsure of whether the release would be gaseous or two-phase, you may want to consider other models or methods to carry out a consequence analysis.

For a hole in the liquid space of a tank, you may use Equation 7-2 below to estimate the release rate. Exhibit B-1 in Appendix B gives the equilibrium vapor pressure in psia for listed toxic gases at 25 °C; this is the pressure required to liquefy the gas at this temperature. You can estimate the gauge pressure in the tank from the equilibrium vapor pressure by subtracting the pressure of the ambient atmosphere (14.7 psi). Exhibit B-1 also gives the Density Factor (DF) for each toxic gas at its boiling point. This factor can be used to estimate the density of the liquefied gas (the density at 25 °C would not be significantly different from the density at the boiling point for most of the listed gases). The equation to estimate the release rate is (see Appendix D, Section D.7.1, for more information):

$$QR = HA \times 6.82 \sqrt{\frac{11.7}{DF^2} \times LH + \frac{669}{DF} \times P_g} \qquad (7\text{-}2)$$

where: QR = Release rate (pounds per minute)
HA = Hole or puncture area (square inches) (from hazard evaluation or best estimate)
DF = Density Factor (listed for each regulated toxic gas in Exhibit B-1, Appendix B; 1/(DF x 0.033) is density in pounds per cubic foot)
LH = Height of liquid column above hole (inches) (from hazard evaluation or best estimate)
P_g = Gauge pressure of the tank pressure (pounds per square inch gauge (psig), from vapor pressure of gas (listed in Exhibit B-1, Appendix B) minus atmospheric pressure (14.7 psi)

This equation gives the rate of release of liquid through the hole. For a gas liquefied under pressure, assume that the released liquid will immediately flash into vapor (or a vapor/aerosol mixture) and the release rate to air is the same as the liquid release rate. This equation gives an estimate of the initial release rate. It will overestimate the overall release rate, because it does not take into account the decrease in the release rate as the pressure in the tank and the height of the liquid in the tank decrease. You may use a computer model or another calculation method if you want a more realistic estimate of the release rate.

For a release from a broken pipe of a gas liquefied under pressure, see equations 7-4 to 7-6 below for liquid releases from pipes. Assume the released liquid flashes into vapor upon release and use the calculated release rate as the release rate to air.

Gases Liquefied by Refrigeration. Gases liquefied by refrigeration alone may be treated as liquids. You may use the methods described in Section 7.2 for estimation of release rates.

Duration of Release

The duration of the release is used in choosing the appropriate generic reference table of distances, as discussed in Chapter 8. (You do not need to consider the duration of the release to use the chemical-specific reference tables.) You may calculate the maximum duration by dividing the quantity in the tank or the quantity that may be released from pipes by your calculated release rate. You may use 60 minutes as a default value for maximum release duration. If you know, and can substantiate, how long it is likely to take to stop the release, you may use that time as the release duration.

7.1.2 Mitigated Releases of Toxic Gases

For gases, passive mitigation may include enclosed spaces, as discussed in Section 3.1.2. Active mitigation for gases, which may be considered in analyzing alternative release scenarios, may include an assortment of techniques including automatic shutoff valves, rapid transfer systems (emergency deinventory), and water/chemical sprays. These mitigation techniques have the effect of reducing either the release rate or the duration of the release, or both.

Active Mitigation

Active Mitigation to Reduce Release Duration. An example of a mitigation technique to reduce the release duration is automatic shutoff valves. If you have an estimate of the rate at which the gas will be released and the time it will take to shut off the release, you may estimate the quantity potentially released (release rate times time). You must be able to substantiate the time it will take to shut off the release. If the release will take place over a period of 10 minutes or more, you may use the release rate to estimate the distance to the toxic endpoint, as discussed in Chapter 8. For releases stopped in less than 10 minutes, multiply the initial release rate by the duration of release to estimate the quantity released, then divide the new quantity by 10 minutes to estimate a mitigated release rate that you may apply to the reference tables described in Chapter 8 to estimate the consequence distance. If the release would be stopped very quickly, you might want to consider other methods that will estimate consequence distances for a puff release.

Active Mitigation to Directly Reduce Release Rate to Air. Examples of mitigation techniques to directly reduce the release rate include scrubbers and flares. Use test data, manufacturer design specifications, or past experience to determine the fractional reduction of the release rate by the mitigation technique. Apply this fraction to the release rate that would have occurred without the mitigation technique. The initial release rate, without mitigation, may be the release rate for the alternative scenario (e.g., a release rate estimated from the equations presented earlier in this section) or the worst-case release rate. The mitigated release rate is:

$$QR_R = (1 - FR) \times QR \qquad (7\text{-}3)$$

where:

QR_R	=	Reduced release rate (pounds per minute)
FR	=	Fractional reduction resulting from mitigation
QR	=	Release rate without mitigation (pounds per minute)

Example 20. Water Spray Mitigation (Hydrogen Fluoride)

A bleeder valve on a hydrogen fluoride (HF) tank opens, releasing 660 pounds per minute of HF. Water sprays are applied almost immediately. Experimental field and laboratory test data indicate that HF vapors could be reduced by 90 percent. The reduced release rate is:

$$QR_R = (1 - 0.9) \times (660 \text{ pounds per minute})$$
$$= 66 \text{ pounds per minute}$$

In estimating the consequence distance for this release scenario, you would need to consider the release both before and after application of the water spray and determine which gives the greatest distance to the endpoint. You need to be able to substantiate the time needed to begin the water spray mitigation.

Passive Mitigation

The same simplified method used for worst-case releases may be used for alternative release scenarios to estimate the release rate to the outside air from a release in an enclosed space. For alternative scenarios, you may use a modified release quantity, if appropriate. You may also adjust the mitigation factor to account for the effects of ventilation, if appropriate for the alternative scenario you have chosen. Use the equations presented in Section 3.1.2 to estimate the release rate to the outside air.

Duration of Release

You should estimate the duration of the release either from your knowledge of the length of time it may take to stop the release (be prepared to substantiate your time estimate) or by dividing the quantity that may be released by your estimated release rate. (You do not need to consider the release duration to use the chemical-specific reference tables of distances.)

7.2 Release Rates for Toxic Liquids

<u>In Section 7.2</u>

- 7.2.1 Methods for estimating the liquid release rate and quantity released for toxic liquids released without mitigation, including:
  - -- Release of toxic liquid from a hole in a tank under atmospheric pressure (including toxic gases liquefied by refrigeration alone),
  - -- Release of toxic liquid from a hole in the liquid space of a pressurized tank (the user is referred to equations provided in the section on toxic gases or in the technical appendix), and
  - -- Release of toxic liquid from a broken pipe.
- 7.2.2 Methods for estimating the liquid release rate and quantity released for toxic liquids released with mitigation measures that reduce the duration of the liquid release or the quantity of liquid released (e.g., automatic shutoff valves).
- 7.2.3 Methods for estimating the evaporation rate of toxic liquids from pools, accounting for:
  - -- Ambient temperature,
  - -- Elevated temperature,
  - -- Diked areas,
  - -- Releases into buildings,
  - -- Active mitigation to reduce the evaporation rate of the liquid,
  - -- Temperatures between 25 °C and 50 °C, and
  - -- Duration of the release.
- 7.2.4 Methods for estimating the evaporation rate for common water solutions of regulated toxic substances and for oleum.

This section describes methods for estimating liquid release rates from tanks and pipes. The released liquid is assumed to form a pool, and the evaporation rate from the pool is estimated as for the worst-case scenario. For the alternative scenario, you may assume the average wind speed in your area in the calculation of evaporation rate, instead of the worst-case wind speed of 1.5 meters per second (3.4 miles per hour). For the reference tables in this guidance, the wind speed for alternative scenarios is assumed to be 3.0 meters per second (6.7 miles per hour).

If you have sufficient information to estimate the quantity of liquid that might be released to an undiked area under an alternative scenario, you may go directly to Section 7.2.3 to estimate the evaporation rate from the pool and the release duration. After you have estimated the evaporation rate and release duration, go to Chapter 8 for instructions on estimating distance to the toxic endpoint.

7.2.1 Liquid Release Rate and Quantity Released for Unmitigated Releases

Release from Tank

Tank under Atmospheric Pressure. If you have a liquid stored in a tank at atmospheric pressure (including gases liquefied by refrigeration alone), you may use the following simple equation to estimate the liquid release rate from a hole in the tank below the liquid level. (See Appendix D, Section D.7.1, for the derivation of this equation.)

$$QR_L = HA \times \sqrt{LH} \times LLF \tag{7-4}$$

where:	QR_L	=	Liquid release rate (pounds per minute)
	HA	=	Hole or puncture area (square inches) (from hazard evaluation or best estimate)
	LH	=	Height of liquid column above hole (inches) (from hazard evaluation or best estimate)
	LLF	=	Liquid Leak Factor incorporating discharge coefficient and liquid density (listed for each toxic liquid in Exhibit B-2, Appendix B).

Remember that this equation only applies to liquids in tanks under atmospheric pressure. This equation will give an overestimate of the release rate, because it does not take into account the decrease in the release rate as the height of the liquid above the hole decreases. You may use a computer model or another calculation method if you want a more realistic estimate of the liquid release rate.

You may estimate the quantity that might be released by multiplying the liquid release rate from the above equation by the time (in minutes) that likely would be needed to stop the release. You should be able to substantiate the time needed to stop the release. Alternatively, you may assume the release would stop when the level of liquid in the tank drops to the level of the hole. You may estimate the quantity of liquid above that level in the tank from the dimensions of the tank, the liquid level at the start of the leak, and the level of the hole. Assume the estimated quantity is released into a pool and use the method and equations in Section 7.2.3 below to determine the evaporation rate of the liquid from the pool and the duration of the release. As discussed in Section 7.2.3, if you find that your estimated evaporation rate is greater than estimated liquid release rate, you should use the liquid release rate as the release rate to air.

Example 21. Liquid Release from Atmospheric Tank (Allyl Alcohol)

You have a tank that contains 20,000 pounds of allyl alcohol at ambient temperature and pressure. A valve on the side of the tank shears, leaving a hole in the tank wall 5 square inches in area. The liquid column is 23 inches above the hole in the tank. From Exhibit B-2, the Liquid Leak Factor for allyl alcohol is 41. Therefore, from Equation 7-4, the liquid release rate is:

$$QR_L = 5 \times (23)^{½} \times 41 = 983 \text{ pounds per minute}$$

It takes 10 minutes to stop the release, so 10 minutes × 983 pounds per minute = 9,830 pounds of allyl alcohol released.

Pressurized Tank. If you have a liquid stored in a tank under pressure, you may estimate a release rate for liquid from a hole in the liquid space of the tank using the equation presented above for gases liquefied under pressure (Equation 7-2 in Section 7.1.1) or the equations in Appendix D, Section D.7.1.

Release from Pipe

If you have a liquid flowing through a pipe at approximately atmospheric pressure, and the pipeline remains at about the same height between the pipe inlet and the pipe break, you can estimate the quantity of liquid released from the flow rate in the pipe and the time it would take to stop the release by multiplying the flow rate by the time. For liquids at atmospheric pressure, assume the liquid is spilled into a pool and use the methods in Section 7.2.3 below to estimate the release rate to air.

For the release of a liquid under pressure from a long pipeline, you may use the equations below (see Appendix D, Section D.7.2 for more information on these equations). These equations apply both to substances that are liquid at ambient conditions and to gases liquefied under pressure. This method does not consider the effects of friction in the pipe. First estimate the initial operational flow velocity of the substance through the pipe using the initial operational flow rate as follows:

$$V_a = \frac{FR \times DF \times 0.033}{A_p} \tag{7-5}$$

where:

V_a	=	Initial operational flow velocity (feet per minute)
FR	=	Initial operational flow rate (pounds per minute)
DF	=	Density Factor (from Exhibit B-2, Appendix B)
A_p	=	Cross-sectional area of pipe (square feet)

You can estimate the cross-sectional area of the pipe from the diameter or radius (half the diameter of the pipe) using the formula for the area of a circle (area = πr^2, where r is the radius).

The release velocity is then calculated based on the initial operational flow, any gravitational acceleration or deceleration effects resulting from changes in the height of the pipeline, and the pressure difference between the pressure in the pipe and atmospheric pressure, using a form of the Bernoulli equation:

$$V_b = 197 \times \sqrt{[28.4 \times (P_T - 14.7) \times DF] + [5.97 \times (Z_a - Z_b)] + [2.58 \times 10^{-5} \times V_a^2]} \quad (7\text{-}6)$$

where:

V_b	=	Release velocity (feet per minute)
P_T	=	Total pressure on liquid in pipe (psia)
DF	=	Density factor, see Exhibit B-1 or Exhibit B-2
Z_a	=	Height of pipeline at inlet (feet)
Z_b	=	Height of pipeline at break (feet)
V_a	=	Operational velocity (feet per minute), calculated from Equation 7-5

Please note that if the height of the pipe at the release point is higher than the initial pipe height, then Z_a-Z_b is negative, and the height term will cause the estimated release velocity to decrease.

The release velocity can then be used to calculate a release rate as follows:

$$QR_L = \frac{V_b \times A_p}{DF \times 0.033} \quad (7\text{-}7)$$

where:

QR_L	=	Release rate (pounds per minute)
V_b	=	Release velocity (feet per minute)
DF	=	Density Factor
A_p	=	Cross-sectional area of pipe (square feet)

You may estimate the quantity released into a pool from the broken pipe by multiplying the liquid release rate (QR_L) from the equation above by the time (in minutes) that likely would be needed to stop the release (or to empty the pipeline). Assume the estimated quantity is released into a pool and use the method and equations described in Section 7.2.3 below to determine the evaporation rate of the liquid from the pool. You must be able to substantiate the time needed to stop the release.

As noted above in Section 7.1.1, for a release from a pipe of gas liquefied under pressure, assume that the released liquid is immediately vaporized, and use the calculated liquid release rate as the release rate to air. If the release duration would be very short (e.g., because of active mitigation measures), determine the total quantity of the release as the release rate times the duration, then estimate a new release rate as the quantity divided by 10. This will give you a release rate that you can use with the 10-minute reference tables of distances in this guidance to estimate a distance to the endpoint.

In the case of very long pipes, release rates from a shear or hole will be lower than the estimates from this method because of pipe roughness and frictional head loss. If friction effects are deemed considerable, an established method for calculating frictional head loss such as the Darcy formula may be used.

7.2.2 Liquid Release Rate and Quantity Released for Mitigated Releases

For alternative release scenarios, you are permitted to take credit for both passive and active mitigation systems, or a combination if both are in place. Active mitigation techniques that reduce the rate of liquid release or the quantity released into the pool are discussed in this section. Active and passive mitigation to reduce the evaporation rate of liquid from a pool are discussed in the next section.

Active Mitigation to Reduce Quantity Released

Examples of mitigation techniques to reduce the quantity released into the pool include automatic shutoff valves and emergency deinventory. You may use the equations in Section 7.2.1 above for calculating liquid release rate, if applicable. Estimate the approximate time needed to stop the release by the mitigation technique (you must be able to justify your estimate). Multiply the release rate times the duration of release to estimate quantity released. Assume the estimated quantity is released into a pool and use the method and equations described in Section 7.2.3 below to determine the evaporation rate of the liquid from the pool. You should also consider mitigation (active or passive) of evaporation from the pool, if applicable, as discussed in Section 7.2.3 below.

Example 22. Mitigated Liquid Release

A bromine injection system suffers a hose failure; the greatly lowered system pressure triggers an automatic shutoff valve within 30 seconds of the release. The flow rate out of the ruptured hose is approximately 330 pounds per minute. Because the release occurred for only 30 seconds (0.5 minutes), the total quantity spilled was 330 x 0.5, or 165 pounds.

7.2.3 Evaporation Rate from Liquid Pool

After you have estimated the quantity of liquid released, assume that the liquid forms a pool and calculate the evaporation rate from the pool as described below. You may account for both passive and active mitigation in estimating the release rate. Passive mitigation may include techniques already discussed in Section 3.2.3 such as dikes and trenches. Active mitigation to reduce the release rate of liquid in pools to the air may include an assortment of techniques including foam or tarp coverings and water or chemical sprays. Some methods of accounting for passive and active mitigation are discussed below.

If the calculated evaporation rate from the pool is greater than the liquid release rate you have estimated from the container, no pool would be formed, and calculating the release rate as the evaporation rate from a pool would not be appropriate. If the pool evaporation rate is greater than the liquid release rate, use the liquid release rate as the release rate to air. Consider this possibility particularly for relatively volatile liquids, gases liquefied by refrigeration, or liquids at elevated temperature that form pools with no mitigation.

Unmitigated

Ambient temperature. For pools with no mitigation, if the liquid is always at ambient temperature, find the Liquid Factor Ambient (LFA) and the Density Factor (DF) in Exhibit B-2 of Appendix B (see Appendix D, Section D.2.2 for the derivation of these factors). If your ambient temperature is between 25 °C and 50 °C, you may use this method to calculate the release rate, and then use the appropriate Temperature Correction Factor from Exhibit B-4, Appendix B, to adjust the release rate, as described below. For gases liquefied by refrigeration, use the Liquid Factor Boiling (LFB) and DF from Exhibit B-1. Calculate the release rate from the following equation for liquids at ambient temperature with no mitigation:

$$QR = QS \times 2.4 \times LFA \times DF \tag{7-8}$$

where:

QR	=	Release rate (pounds per minute)
QS	=	Quantity released (pounds)
2.4	=	Wind speed factor = $3.0^{0.78}$, where 3.0 meters per second (6.7 miles per hour) is the wind speed for the alternative scenario for purposes of this guidance
LFA	=	Liquid Factor Ambient
DF	=	Density Factor

This method assumes that the total quantity of liquid released spreads out to form a pool one centimeter in depth; it does not take into account evaporation as the liquid is released.

Example 23. Evaporation from Pool Formed by Liquid Released from Hole in Tank (Allyl Alcohol)

In Example 21, 9,830 pounds of allyl alcohol were estimated to be released from a hole in a tank. From Exhibit B-2, the Density Factor for allyl alcohol is 0.58, and the Liquid Factor Ambient is 0.0046. Assuming that the liquid is not released into a diked area or inside a building, the evaporation rate from the pool of allyl alcohol, from Equation 7-8, is:

$$QR = 9{,}830 \times 2.4 \times 0.0046 \times 0.58 = 63 \text{ pounds per minute}$$

Elevated temperature. For pools with no mitigation, if the liquid is at an elevated temperature (above 50 °C or at or close to its boiling point), find the Liquid Factor Boiling (LFB) and the Density Factor (DF) in Exhibit B-2 of Appendix B (see Appendix D, Section D.2.2, for the derivation of these factors). For liquids at temperatures between 25 °C and 50 °C, you may use the method above for ambient temperature and apply the appropriate Temperature Correction Factor from Appendix B, Exhibit B-4, to the result, as discussed below. For liquids above 50 °C, or close to their boiling points, or with no Temperature Correction Factors available, calculate the release rate of the liquid from the following equation:

$$QR = QS \times 2.4 \times LFB \times DF \tag{7-9}$$

where:

QR	=	Release rate (pounds per minute)
QS	=	Quantity released (pounds)
2.4	=	Wind speed factor = $3.0^{0.78}$, where 3.0 meters per second (6.7 miles per hour) is the wind speed for the alternative scenario for purposes of this guidance
LFB	=	Liquid Factor Boiling
DF	=	Density Factor

Mitigated

Diked Areas. If the toxic liquid will be released into an area where it will be contained by dikes, compare the diked area to the maximum area of the pool that could be formed, as described in Section 3.2.3 (see Equation 3-6). Also verify that the quantity spilled will be totally contained by the dikes. The smaller of the two areas should be used in determination of the evaporation rate. If the maximum area of the pool is smaller than the diked area, calculate the release rate as described for pools with no mitigation (above). If the diked area is smaller, and the spill will be totally contained, go to Exhibit B-2 in Appendix B to find the Liquid Factor Ambient (LFA), if the liquid is at ambient temperature, or the Liquid Factor Boiling (LFB), if the liquid is at an elevated temperature. For temperatures between 25 °C and 50 °C, you may use the appropriate Temperature Correction Factor from Exhibit B-4, Appendix B, to adjust the release rate, as described below. For gases liquefied by refrigeration, use the LFB. Calculate the release rate from the diked area as follows for liquids at ambient temperature:

$$QR = 2.4 \times LFA \times A \qquad (7\text{-}10)$$

or, for liquids at elevated temperatures or gases liquefied by refrigeration alone:

$$QR = 2.4 \times LFB \times A \qquad (7\text{-}11)$$

where:

QR	=	Release rate (pounds per minute)
2.4	=	Wind speed factor = $3.0^{0.78}$, where 3.0 meters per second (6.7 miles per hour) is the wind speed for the alternative scenario for purposes of this guidance
LFA	=	Liquid Factor Ambient (listed in Exhibit B-2, Appendix B)
LFB	=	Liquid Factor Boiling (listed in Exhibit B-1 or B-2, Appendix B)
A	=	Diked area (square feet)

Releases Into Buildings. If a toxic liquid is released inside a building, compare the area of the building floor or any diked area that would contain the spill to the maximum area of the pool that could be formed; the smaller of the two areas should be used in determining the evaporation rate, as for the worst-case scenario. The maximum area of the pool is determined from Equation 3-6 in Section 3.2.3 for releases into diked areas. The area of the building floor is the length times width of the floor (in feet) (Equation 3-9).

If the floor area or diked area is smaller than the maximum pool size, estimate the outdoor evaporation rate from a pool the size of the floor area or diked area from Equation 7-10. If the maximum pool area is smaller, estimate the outdoor evaporation rate from a pool of maximum size from Equation 7-8.

Estimate the rate of release of the toxic vapor from the building as five percent of the calculated outdoor evaporation rate (multiply your evaporation rate by 0.05). See Appendix D, Section D.2.4 for more information on releases into buildings.

Active Mitigation to Reduce Evaporation Rate. Examples of active mitigation techniques to reduce the evaporation rate from the pool include water sprays and foam or tarp covering. Use test data, manufacturer design specifications, or past experience to determine the fractional reduction of the release rate by the mitigation technique. Apply this fraction to the release rate (evaporation rate from the pool) that would have occurred without the mitigation technique, as follows:

$$QR_{RV} = (1 - FR) \times QR \tag{7-12}$$

where:

QR_{RV}	=	Reduced evaporation rate (release rate to air) from pool (pounds per minute)
FR	=	Fractional reduction resulting from mitigation
QR	=	Evaporation rate from pool without mitigation (pounds per minute)

Temperature Corrections for Liquids at Temperatures between 25 and 50 °C

If your liquid is at a temperature between 25 °C (77 °F) and 50 °C (122 °F), you may use the appropriate Temperature Correction Factor (TCF) from Exhibit B-4, Appendix B, to calculate a corrected release rate. Calculate the release rate (QR) of the liquid at 25 °C (77 °F) as described above for unmitigated releases or releases in diked areas and multiply the release rate by the appropriate TCF as described in Section 3.2.5.

Evaporation Rate Compared to Liquid Release Rate

If you estimated the quantity of liquid in the pool based on an estimated liquid release rate from a hole in a container or pipe, as discussed in Sections 7.2.1 and 7.2.2, compare the evaporation rate with the liquid release rate. If the evaporation rate from the pool is greater than the liquid release rate, use the liquid release rate as the release rate to air.

Duration of Release

After you have estimated a release rate as described above, determine the duration of the vapor release from the pool (the time it will take for the liquid pool to evaporate completely). To estimate the time in minutes, divide the total quantity released (in pounds) by the release rate (in pounds per minute) (see Equation 3-5 in Section 3.2.2). If you are using the liquid release rate as the release rate to air, as discussed in the preceding paragraph, estimate a liquid release duration as discussed in Section 7.2.1 or 7.2.2. The duration could be the time it would take to stop the release or the time it would take to empty the tank or to release all the liquid above the level of the leak. If you have corrected the release rate for a temperatures above 25 °C, use the corrected release rate to estimate the duration.

7.2.4 Common Water Solutions and Oleum

You may use the methods described above in Sections 7.2.1, 7.2.2, and 7.2.3 for pure liquids to estimate the quantity of a solution of a toxic substance or oleum that may be spilled into a pool. LFA, DF, and LLF values for several concentrations of ammonia, formaldehyde, hydrochloric acid, hydrofluoric acid, and nitric acid in water solution and for oleum are listed in Appendix B, Exhibit B-3. The LFA for a wind speed of 3.0 meters per second (6.7 miles per hour) should be used in the release rate calculations for alternative scenarios for pools of solutions at ambient temperature.

For unmitigated releases or releases with passive mitigation, follow the instructions in Section 7.2.3. If active mitigation measures are in place, you may estimate a reduced release rate from the instructions on active mitigation in Section 7.2.2. Use the total quantity of the solution as the quantity released from the vessel or pipeline (QS) in carrying out the calculation of the release rate to the atmosphere.

If the solution is at an elevated temperature, see Section 3.3. As discussed in Section 3.3, you may treat the release of the substance in solution as a release of the pure substance. Alternatively, if you have vapor pressure data for the solution at the release temperature, you may estimate the release rate from the equations in Appendix D, Sections D.2.1 and D.2.2.

If you estimated the quantity of solution in the pool based on an estimated liquid release rate from a hole in a container or pipe, as discussed in Sections 7.2.1 and 7.2.2, compare the evaporation rate with the liquid release rate. If the evaporation rate from the pool is greater than the liquid release rate, use the liquid release rate as the release rate to air.

8 ESTIMATION OF DISTANCE TO THE ENDPOINT FOR ALTERNATIVE SCENARIOS FOR TOXIC SUBSTANCES

In Chapter 8

- Reference tables of distances for alternative releases, including:
  - -- Generic reference tables (Exhibit 4), and
  - -- Chemical-specific reference tables (Exhibit 5).
- Considerations include:
  - -- Gas density (neutrally buoyant or dense),
  - -- Duration of release (10 minutes or 60 minutes),
  - -- Topography (rural or urban).

For estimating consequence distances for alternative scenarios for toxic substances, this guidance provides four generic reference tables for neutrally buoyant gases and vapors and four for dense gases. The generic reference tables of distances (Reference Tables 14-21) are found at the end of Chapter 10. The generic tables and the conditions for which each table is applicable are described in Exhibit 4. Four chemical-specific tables also are provided for ammonia, chlorine, and sulfur dioxide. The chemical-specific reference tables follow the generic reference tables at the end of Chapter 10. These tables, and the applicable conditions, are described in Exhibit 5.

All the reference tables of distances for alternative scenarios were developed assuming D stability and a wind speed of 3.0 meters per second (6.7 miles per hour) as representative of likely conditions for many sites. Many wind speed and atmospheric stability combinations may be possible at different times in different parts of the country. If D stability and 3.0 meters per second are not reasonable conditions for your site, you may want to use other methods to estimate distances.

For simplicity, this guidance assumes ground level releases. This guidance, therefore, may overestimate the consequence distance if your alternative scenario involves a release above ground level, particularly for neutrally buoyant gases and vapors. If you want to assume an elevated release, you may want to consider other methods to determine the consequence distance.

The generic reference tables should be used for all toxic substances other than ammonia, chlorine, and sulfur dioxide. To use the generic reference tables, you need to consider the release rates estimated for gases and evaporation from liquid pools and the duration of the release. For the alternative scenarios, the duration of toxic gas releases may be longer than the 10 minutes assumed for the worst-case analysis for gases. You need to determine the appropriate toxic endpoint and whether the gas or vapor is neutrally buoyant or dense, using the tables in Appendix B and considering the conditions of the release. You may interpolate between entries in the reference tables.

Exhibit 4
Generic Reference Tables of Distances for Alternative Scenarios

Applicable Conditions			Reference Table Number
Gas or Vapor Density	**Topography**	**Release Duration (minutes)**	
Neutrally buoyant	Rural	10	14
		60	15
	Urban	10	16
		60	17
Dense	Rural	10	18
		60	19
	Urban	10	20
		60	21

Exhibit 5
Chemical-Specific Reference Tables of Distances for Alternative Scenarios

Substance	Conditions of Release			Reference Table Number
	Gas or Vapor Density	**Release Duration (minutes)**	**Topography**	
Anhydrous ammonia liquefied under pressure	Dense	10-60	Rural, urban	22
Non-liquefied ammonia, ammonia liquefied by refrigeration, or aqueous ammonia	Neutrally buoyant	10-60	Rural, urban	23
Chlorine	Dense	10-60	Rural, urban	24
Sulfur dioxide (anhydrous)	Dense	10-60	Rural, urban	25

Note the following concerning the use of the chemical-specific reference tables for ammonia, chlorine, and sulfur dioxide:

- The table for anhydrous ammonia (Reference Table 22) applies only to flashing releases of ammonia liquefied under pressure. Use Table 23 for release of ammonia as a gas (e.g., evaporation from a pool or release from the vapor space of a tank).

- You may use these tables for releases of any duration.

To use the reference tables of distances, follow these steps:

For Regulated Toxic Substances Other than Ammonia, Chlorine, and Sulfur Dioxide

- Find the toxic endpoint for the substance in Appendix B (Exhibit B-1 for toxic gases or Exhibit B-2 for toxic liquids).

- Determine whether the table for neutrally buoyant or dense gases and vapors is appropriate from Appendix B (Exhibit B-1 for toxic gases or Exhibit B-2, column for alternative case, for toxic liquids). A toxic gas that is lighter than air may behave as a dense gas upon release if it is liquefied under pressure, because the released gas may be mixed with liquid droplets, or if it is cold. Consider the state of the released gas when you decide which table is appropriate.

- Determine whether the table for rural or urban conditions is appropriate.

  -- Use the rural table if your site is in an open area with few obstructions.

  -- Use the urban table if your site is in an urban or obstructed area.

- Determine whether the 10-minute table or the 60-minute table is appropriate.

  -- Use the 10-minute table for releases from evaporating pools of common water solutions and of oleum.

  -- If you estimated the release duration for gas release or pool evaporation to be 10 minutes or less, use the 10-minute table.

  -- If you estimated the release duration for gas release or pool evaporation to be more than 10 minutes, use the 60-minute table.

<u>Neutrally Buoyant Gases or Vapors</u>

- If Exhibit B-1 or B-2 indicates the gas or vapor should be considered neutrally buoyant, and other factors would not cause the gas or vapor to behave as a dense gas, divide the estimated release rate (pounds per minute) by the toxic endpoint (milligrams per liter).

- Find the range of release rate/toxic endpoint values that includes your calculated release rate/toxic endpoint in the first column of the appropriate table (Reference Table 14, 15, 16, or 17), then find the corresponding distance to the right.

  Dense Gases or Vapors

- If Exhibit B-1 or B-2 or consideration of other relevant factors indicates the substance should be considered a dense gas or vapor (heavier than air), find the distance in the appropriate table (Reference Table 18, 19, 20, or 21) as follows;

  -- Find the toxic endpoint closest to that of the substance by reading across the top of the table. If the endpoint of the substance is halfway between two values on the table, choose the value on the table that is smaller (to the left). Otherwise, choose the closest value to the right or the left.

  -- Find the release rate closest to the release rate estimated for the substance at the left of the table. If the calculated release rate is halfway between two values on the table, choose the release rate that is larger (farther down on the table). Otherwise, choose the closest value (up or down on the table).

  -- Read across from the release rate and down from the endpoint to find the distance corresponding to the toxic endpoint and release rate for your substance.

For Ammonia, Chlorine, or Sulfur Dioxide

- Find the appropriate chemical-specific table for your substance (see the descriptions of Reference Tables 22-25 in Exhibit 5).

  -- If you have ammonia liquefied by refrigeration alone, you may use Reference Table 23, even if the duration of the release is greater than 10 minutes.

  -- If you have chlorine or sulfur dioxide liquefied by refrigeration alone, you may use the chemical-specific reference tables, even if the duration of the release is greater than 10 minutes.

- Determine whether rural or urban topography is applicable to your site.

  -- Use the rural column in the reference table if your site is in an open area with few obstructions.

  -- Use the urban column if your site is in an urban or obstructed area.

- Estimate the consequence distance as follows:

  -- In the left-hand column of the table, find the release rate closest to your calculated release rate.

-- Read the corresponding distance from the appropriate column (urban or rural) to the right.

The development of the generic reference tables is discussed in Appendix D, Sections D.4.1 and D.4.2. The development of the chemical-specific reference tables is discussed in industry-specific risk management program guidance documents and a backup information document that are cited in Section D.4.3. If you think the results of the method presented here overstate the potential consequences of a your alternative release scenario, you may choose to use other methods or models that take additional site-specific factors into account.

Examples 24 and 25 below include the results of modeling using two other models, ALOHA and WHAZAN, for comparison with the results of the methods presented in this guidance. Appendix D, Section D.4.5 provides additional information on this modeling.

Example 24. Gas Release of Chlorine

Assume that you calculated a release rate of 500 pounds per minute of chlorine from a tank. A chemical-specific table is provided for chlorine, so you do not need to consult Appendix B for information on chlorine. The topography of your site is urban. For a release of chlorine under average meteorology (D stability and 3 meters per second wind speed), go to Reference Table 24. The estimated release rate of 500 pounds per minute, with urban topography, corresponds to a consequence distance of 0.4 miles.

Additional Modeling for Comparison

The ALOHA model gave a distance of 3.0 miles to the endpoint, using the same assumptions.

The WHAZAN model gave a distance of 3.2 miles to the endpoint, using the same assumptions and the dense cloud dispersion model.

Example 25. Allyl Alcohol Evaporating from Pool

In Example 23, the evaporation rate of allyl alcohol from a pool was calculated as 63 pounds per minute. The total quantity in the pool was estimated as 9,830 pounds; therefore, the pool would evaporate in 9,830/63 or 156 minutes. You would use a 60-minute reference table to estimate the distance to the endpoint. From Exhibit B-2 in Appendix B, the toxic endpoint for allyl alcohol is 0.036 mg/L, and the appropriate reference table for the alternative scenario analysis is a neutrally buoyant plume table. To find the distance from the neutrally buoyant plume tables, you need the release rate divided by the endpoint. In this case, it is 63/0.036, or 1,750. Assuming the release takes place in a rural location, you use Reference Table 15, applicable to neutrally buoyant plumes, 60-minute releases, and rural conditions. From this table, you estimate the distance as 0.4 mile.

Additional Modeling for Comparison

The ALOHA model gave a distance of 0.7 mile to the endpoint for a release rate of 63 pounds per minute, using the same assumptions and the dense gas model.

The WHAZAN model gave a distance of 0.5 mile to the endpoint for a release rate of 63 pounds per minute, using the same assumptions and the buoyant plume dispersion model.

9 ESTIMATION OF RELEASE RATES FOR ALTERNATIVE SCENARIOS FOR FLAMMABLE SUBSTANCES

In Chapter 9

- Methods to estimate a release rate to air for a flammable gas (9.1) or liquid (9.2).

9.1 Flammable Gases

Gaseous Release from Tank or Pipe

An alternative scenario for a release of a flammable gas may involve a leak from a vessel or piping. To estimate a release rate for flammable gases from hole size and storage conditions, you may use the method described above in Section 7.1.1 for toxic gases. This release rate may be used to determine the dispersion distance to the lower flammability limit (LFL), as described in Section 10.1. Exhibit C-2 in Appendix C includes Gas Factors (GF) that may be used in carrying out the calculations for each of the regulated flammable gases.

Example 26. Release Rate of Flammable Gas from Hole in Tank (Ethylene)

A pipe tears off a tank containing ethylene. The pipe is in the vapor space of the tank. The release rate from the hole can be estimated from Equation 7-1 in Section 7.1. You estimate that the pipe would leave a hole with an area (HA) of 5 square inches. The temperature inside the tank (T_t, absolute temperature, Kelvin) is 282 K, 9 C, and the square root of the temperature is 16.8. The pressure in the tank (P_t) is approximately 728 pounds per square inch absolute (psia). From Exhibit C-2, Appendix C, the gas factor (GF) for ethylene is 18. From Equation 7-1, the release rate (QR) is:

$$QR = 5 \times 728 \times (1/16.8) \times 18 = 3{,}900 \text{ pounds per minute}$$

Gases Liquefied Under Pressure

A vapor cloud fire is a possible result of a release of a gas liquefied under pressure. You may use the methods described in Section 7.1.1 for toxic gases liquefied under pressure to estimate the release rate from a hole in a tank for a flammable gas liquefied under pressure. The estimated release rate may be used to estimate the dispersion distance to the LFL for a vapor cloud fire.

Flammable gases that are liquefied under pressure may be released very rapidly, with partial vaporization of the liquefied gas and possible aerosol formation. Section 10.4 presents a method for estimating the consequences of a vapor cloud explosion from such a release of a gas liquefied under pressure.

Gases Liquefied by Refrigeration

Flammable gases liquefied by refrigeration alone can be treated as liquids for the alternative scenario analysis, as discussed in Section 9.2 and Section 10.2, below.

9.2 Flammable Liquids

You may estimate a release rate for flammable liquids by estimating the evaporation rate from a pool. Release rates also can be estimated for flammable gases liquefied by refrigeration alone by this method, if the liquefied gas is likely to form a pool upon release. You first need to estimate the quantity in the pool.

You may use the method discussed in Section 7.2.1 to estimate a rate of liquid release for flammable liquids into a pool from a hole in a tank or from a pipe shear. Exhibit C-3 in Appendix C includes liquid leak factors (LLF) for calculating release rate from a hole. Note that the LLF is appropriate only for atmospheric tanks. LLF values are not provided for liquefied flammable gases; you will need to estimate the quantity in the pool from other information for liquefied flammable gases.

Once you have an estimate of the quantity of flammable liquid in a pool, you may use the methods presented in Section 7.2.3 to estimate the evaporation rate from the pool. Liquid factors at ambient and boiling temperature (LFA and LFB) for liquids for the calculation are listed in Exhibit C-3 in Appendix C, and LFBs for liquefied gases are listed in Exhibit C-2. Both passive and active mitigation measures (discussed in Sections 7.2.2 and 7.2.3) may be taken into account. You do not need to estimate the duration of the release, because this information is not used to estimate distance to the LFL, as discussed in the next chapter.

As for toxic liquids, if the rate of evaporation of the liquid from the pool is greater than the rate of release of the liquid from the container, you should use the liquid release rate, not the pool evaporation rate, as the rate of release to the air. You should expect rapid evaporation rates for liquefied flammable gases from a pool. All of the regulated flammable liquids are volatile, so the evaporation rate from a pool may be expected to be relatively high, particularly without mitigation.

10 ESTIMATION OF DISTANCE TO THE ENDPOINT FOR ALTERNATIVE SCENARIOS FOR FLAMMABLE SUBSTANCES

In Chapter 10

- 10.1 Method to estimate the dispersion distance to the LFL for vapor cloud fires.
- 10.2 Method to estimate the distance to the heat radiation endpoint for a pool fire involving a flammable liquid, based on the pool area and factors provided in the appendix.
- 10.3 Method to estimate the distance to the heat radiation endpoint for a fireball from a BLEVE, using a reference table of distances.
- 10.4 Alternative scenario analysis for vapor cloud explosions, using less conservative assumptions than for worst-case vapor cloud explosions.

10.1 Vapor Cloud Fires

The distance to the LFL represents the maximum distance at which the radiant heat effects of a vapor cloud fire might have serious consequences. Exhibit C-2, Appendix C, provides LCL data (in volume percent and milligrams per liter) for listed flammable gases; Exhibit C-3 provides these data for flammable liquids. This guidance provides reference tables for the alternative scenario conditions assumed in this guidance (D stability and wind speed 3.0 meters per second, ground level releases) for estimating the distance to the LCL. Release rate is the primary factor for determining distance to the flammable endpoint. Because the methods used in this guidance assumes that the vapor cloud release is in a steady state and that vapor cloud fires are nearly instantaneous events, release duration is not a critical factor for estimating vapor cloud fire distances. Thus, the reference tables for flammable substances are not broken out separately by release duration (e.g., 10 minutes, 60 minutes). The development of these tables is discussed further in Appendix D, Section D.4. The reference tables for flammable substances (Reference Tables 26-29 at the end of Chapter 10) are listed in Exhibit 6.

To use the reference tables of distances to find the distance to the LFL from the release rate, follow these steps:

- Find the LFL endpoint for the substance in Appendix C (Exhibit C-2 for flammable gases or Exhibit C-3 for flammable liquids).
- Determine from Appendix C whether the table for neutrally buoyant or dense gases and vapors is appropriate (Exhibit C-2 for flammable gases or Exhibit C-3 for flammable liquids). A gas that is lighter than air may behave as a dense gas upon release if it is liquefied under pressure, because the released gas may be mixed with liquid droplets, or if it is cold. Consider the state of the released gas when you decide which table is appropriate.

- Determine whether the table for rural or urban conditions is appropriate.
  - -- Use the rural table if your site is in an open area with few obstructions.
  - -- Use the urban table if your site is in an urban or obstructed area.

Exhibit 6
Reference Tables of Distances for Vapor Cloud Fires of Flammable Substances

Applicable Conditions			Reference Table Number
Gas or Vapor Density	**Topography**	**Release Duration (minutes)**	
Neutrally buoyant	Rural	10 - 60	26
	Urban	10 - 60	27
Dense	Rural	10 - 60	28
	Urban	10 - 60	29

Neutrally Buoyant Gases or Vapors

- If Exhibit C-2 or C-3 indicates the gas or vapor should be considered neutrally buoyant, and other factors would not cause the gas or vapor to behave as a dense gas, divide the estimated release rate (pounds per minute) by the LFL endpoint (milligrams per liter).
- Find the range of release rate/LFL values that includes your calculated release rate/LFL in the first column of the appropriate table (Reference Table 26 or 27), then find the corresponding distance to the right.

Dense Gases or Vapors

- If Exhibit C-2 or C-3 or consideration of other relevant factors indicates the substance should be considered a dense gas or vapor (heavier than air), find the distance in the appropriate table (Reference Table 28 or 29) as follows:
  - -- Find the LFL closest to that of the substance by reading across the top of the table. If the LFL of the substance is halfway between two values on the table, choose the value on the table that is smaller (to the left). Otherwise, choose the closest value to the right or the left.
  - -- Find the release rate closest to the release rate estimated for the substance at the left of the table. If the calculated release rate is halfway between two values on the

table, choose the release rate that is larger (farther down on the table). Otherwise, choose the closest value (up or down on the table).

-- Read across from the release rate and down from the LFL to find the distance corresponding to the LFL and release rate for your substance.

Example 27. Flammable Gas Release (Ethylene)

In Example 26, you estimated a release rate for ethylene from a hole in a tank of 3,900 pounds per minute. You want to estimate the distance to the LFL for a vapor cloud fire resulting from this release.

From Exhibit C-2, Appendix C, the LFL for ethylene is 31 mg/L, and the appropriate table for distance estimation is a neutrally buoyant gas table for flammable substances. Your site is in a rural area, so you would use Reference Table 26.

To use the neutrally buoyant gas tables, you need to calculate release rate/endpoint. In this case, release rate/LFL = 3,900/31 or 126. On Reference Table 26, 126 falls in the range of release rate/LFL values corresponding to 0.2 miles.

Example 28. Vapor Cloud Fire from Evaporating Pool of Flammable Liquid

You have a tank containing 20,000 pounds of ethyl ether. A likely scenario for a release might be shearing of a pipe from the tank, with the released liquid forming a pool. You want to estimate the consequences of a vapor cloud fire that might result from evaporation of the pool and ignition of the vapor.

You first need to estimate the rate of release of the liquid from the tank. You can do this using Equation 7-4, Section 7.2.1. For this calculation, you need the area of the hole that would result from shearing the pipe (HA), the height of the liquid in the tank above the hole (LH), and the liquid leak factor (LLF) for ethyl ether, from Exhibit C-3 in Appendix C. The pipe diameter is 2 inches, so the cross sectional area of the hole would be 3.1 square inches. You estimate that the pipe is 2 feet, or 24 inches, below the level of the liquid when the tank is full. The square root of LH (24 inches) is 4.9. LLF for ethyl ether is 34. From Equation 7-4, the rate of release of the liquid from the hole is calculated as:

$$\begin{aligned} QR_L &= 3.1 \times 4.9 \times 34 \\ &= 520 \text{ pounds per minute} \end{aligned}$$

You estimate that the release of the liquid could be stopped in about 10 minutes. In 10 minutes, 10 × 520, or 5,200 pounds, would be released.

The liquid would be released into an area without dikes. To estimate the evaporation rate from the pool formed by the released liquid, you use Equation 7-8 from Section 7.2.3. To carry out the calculation, you need the Liquid Factor Ambient (LFA) and the Density Factor (DF) for ethyl ether. From Exhibit C-3, Appendix C, LFA for ethyl ether is 0.11 and DF is 0.69. The release rate to air is:

$$\begin{aligned} QR &= 5{,}200 \times 2.4 \times 0.11 \times 0.69 \\ &= 950 \text{ pounds per minute} \end{aligned}$$

The evaporation rate from the pool is greater than the estimated liquid release rate; therefore, you use the liquid release rate of 520 pounds per minute as the release rate to air. To estimate the maximum distance at which people in the area of the vapor cloud could suffer serious injury, estimate the distance to the lower flammability limit (LFL) (in milligrams per liter) for ethyl ether, from the appropriate reference table. From Exhibit C-3, Appendix C, LFL for ethyl ether is 57 mg/L, and the appropriate reference table is a dense gas table. Your site is in a rural area with few obstructions, so you use Reference Table 28.

From Reference Table 28, the closest LFL is 60 mg/L. The lowest release rate on the table is 1,500 pounds per minute, which is higher than the evaporation rate estimated for the pool of ethyl ether. For a release rate less than 1,500 pounds per minute, the distance to the LFL is less than 0.1 miles.

10.2 Pool Fires

Pool fires may be considered as potential alternative scenarios for flammable liquids, including gases liquefied by refrigeration alone. You may find, however, that other scenarios will give a greater distance to the endpoint and, therefore, may be more appropriate as alternative scenarios. A "Pool Fire Factor" (PFF) has been derived for each of the regulated flammable liquids and most of the flammable gases to aid in the consequence analysis. The derivation of these factors is discussed in Appendix D, Section D.9. The PFF, listed in Appendix C, Exhibit C-2 for flammable gases and C-3 for flammable liquids, may be used to estimate a distance from the center of a pool fire where people could potentially receive second degree burns from a 40-second exposure. The heat radiation endpoint for this analysis is 5 kilowatts per square meter (kW/m^2). Ambient temperature is assumed to be 25 °C (77 °F) for calculation of the PFF for flammable liquids.

To estimate a distance using the PFF, you first need to estimate the size of the pool, in square feet, that might be formed by the release of a flammable substance. You may use the methods described above for toxic liquids to estimate pool size. Density factors (DF) for the estimation of pool size in undiked areas may be found for flammable gases and flammable liquids in Exhibits C-2 and C-3 of Appendix C. For flammable gases, the DF is based on the density at the boiling point. You may want to consider whether the released substance may evaporate too quickly to form a pool of the maximum size, particularly for liquefied gases.

Distances may be estimated from the PFF and the pool area as follows:

$$d = PFF \times \sqrt{A} \qquad (10\text{-}1)$$

where:

d	=	Distance (feet)
PFF	=	Pool Fire Factor (listed for each flammable substance in Appendix C, Exhibits C-2 and C-3)
A	=	Pool area (square feet)

Example 29. Pool Fire of Flammable Liquid

For a tank containing 20,000 pounds of ethyl ether, you want to estimate the consequences of a pool fire. You estimate that 15,000 pounds would be released into an area without dikes, forming a pool. Assuming the liquid spreads to a depth of 1 centimeter (0.39 inches), you estimate the area of the pool formed from Equation 3-6, Section 3.2.3. For this calculation, you need the density factor (DF) for ethyl ether; from Exhibit C-3, Appendix C, DF for ethyl ether is 0.69. From Equation 3-6, the area of the pool is:

$$A = 15{,}000 \times 0.69 = 10{,}400 \text{ square feet}$$

You can use Equation 10-1 to estimate the distance from the center of the burning pool where the heat radiation level would reach 5 kW/m^2. For the calculation, you need the square root of the pool area (A) and the pool fire factor (PFF) for ethyl ether. The square root of A, 10,400 square feet, is 102 feet. From Exhibit C-3, Appendix C, PFF for ethyl ether is 4.3. From Equation 10-1, the distance (d) to 5 kW/m^2 is:

$$d = 4.3 \times 102 = 440 \text{ feet (about 0.08 miles)}$$

If you have a gas that is liquefied under pressure or under a combination of pressure and refrigeration, a pool fire is probably not an appropriate alternative scenario. A fire or explosion involving the flammable gas that is released to the air by a sudden release of pressure is likely to have the potential for serious effects at a greater distance than a pool fire (e.g., see the methods for analysis of BLEVEs and vapor cloud explosions in Sections 10.3 and 10.4 below, or see Appendix A for references that provide more information on consequence analysis for fires and explosions).

10.3 BLEVEs

If a fireball from a BLEVE is a potential release scenario at your site, you may use Reference Table 30 to estimate the distance to a potentially harmful radiant heat level. The table shows distances for a range of quantities to the radiant heat level that potentially could cause second degree burns to a person exposed for the duration of the fire. The quantity you use should be the total quantity in a tank that might be involved in a BLEVE. The equations used to derive this table of distances are presented in Appendix D, Section D.10. If you prefer, you may use the equations to estimate a distance for BLEVEs, or you may use a different calculation method or model.

10.4 Vapor Cloud Explosion

If you have the potential at your site for the rapid release of a large quantity of a flammable vapor, particularly into a congested area, a vapor cloud explosion may be an appropriate alternative release scenario. For the consequence analysis, you may use the same methods as for the worst case to estimate consequence distances to an overpressure endpoint of 1 psi (see Section 5.1 and the equation in Appendix C). Instead of assuming the total quantity of flammable substance released is in the vapor cloud, you may estimate a smaller quantity in the cloud. You could base your estimate of the quantity in the cloud on the release rate estimated as described above for gases and liquids multiplied by the time required to stop the release.

To estimate the quantity in the cloud for a gas liquefied under pressure (not refrigerated), you may use the equation below. This equation incorporates a "flash fraction factor" (FFF), listed in Appendix C, Exhibit C-2 for regulated flammable gases, to estimate the quantity that could be immediately flashed into vapor upon release. A factor of two is included to estimate the quantity that might be carried along as spray or aerosol. See Appendix D, Section D.11 for the derivation of this equation. The equation is:

$$QF = FFF \times QS \times 2 \qquad \textbf{(10-2)}$$

where:	QF	=	Quantity flashed into vapor plus aerosol (pounds) (cannot be larger than QS)
	FFF	=	Flash fraction factor (unitless) (listed in Appendix C, Exhibit C-2) (must be less than 1)
	QS	=	Quantity spilled (pounds)
	2	=	Factor to account for spray and aerosol

For derivation of the FFF, the temperature of the stored gas was assumed to be 25 °C (77 °F) (except as noted in Exhibit C-2). You may estimate the flash fraction under other conditions using the equation presented in Appendix D, Section D.11.

You may estimate the distance to 1 psi for a vapor cloud explosion from the quantity in the cloud using Reference Table 13 (at the end of the worst-case analysis discussion) or from Equation C-1 in Appendix C. For the alternative scenario analysis, you may use a yield factor of 3 percent, instead of the yield factor of 10 percent used in the worst-case analysis. As discussed in Appendix D, Section D.11, the yield factor of 3 percent is representative of more likely events, based on data from past vapor cloud explosions. If you use the equation in Appendix C, use 0.03 instead of 0.1 in the calculation. If you use Reference Table 13, you can incorporate the lower yield factor by multiplying the distance you read from Reference Table 13 by 0.67.

Example 30. Vapor Cloud Explosion (Propane)

You have a tank containing 50,000 pounds of propane liquefied under pressure at ambient temperature. You want to estimate the consequence distance for a vapor cloud explosion resulting from rupture of the tank.

You use Equation 10-2 to estimate the quantity that might be released to form a cloud. You base the calculation on the entire contents of the tank (QS = 50,000 pounds). From Exhibit C-2 of Appendix C, the Flash Fraction Factor (FFF) for propane is 0.38. From Equation 10-2, the quantity flashed into vapor, plus the quantity that might be carried along as aerosol, (QF) is:

$$QF = 0.38 \times 50{,}000 \times 2 = 38{,}000 \text{ pounds}$$

You assume 38,000 pounds of propane is in the flammable part of the vapor cloud. This quantity falls between 20,000 pounds and 50,000 pounds in Reference Table 13; 50,000 pounds is the quantity closest to your quantity. From the table, the distance to 1 psi overpressure is 0.3 mile for 50,000 pounds of propane for a 10 percent yield factor. To change the yield factor to 3 percent, you multiply this distance by 0.67; then the distance becomes 0.2 mile.

Reference Table 14

Neutrally Buoyant Plume Distances to Toxic Endpoint for Release Rate Divided by Endpoint
10-Minute Release, Rural Conditions, D Stability, Wind Speed 3.0 Meters per Second

Release Rate/Endpoint [(lbs/min)/(mg/L)]	Distance to Endpoint (miles)
0 - 64	0.1
64 - 510	0.2
510 - 1,300	0.3
1,300 - 2,300	0.4
2,300 - 4,100	0.6
4,100 - 6,300	0.8
6,300 - 8,800	1.0
8,800 - 12,000	1.2
12,000 - 16,000	1.4
16,000 - 19,000	1.6
19,000 - 22,000	1.8
22,000 - 26,000	2.0
26,000 - 30,000	2.2
30,000 - 36,000	2.4
36,000 - 42,000	2.6
42,000 - 47,000	2.8
47,000 - 54,000	3.0
54,000 - 60,000	3.2
60,000 - 70,000	3.4
70,000 - 78,000	3.6
78,000 - 87,000	3.8
87,000 - 97,000	4.0
97,000 - 110,000	4.2
110,000 - 120,000	4.4
120,000 - 130,000	4.6

Release Rate/Endpoint [(lbs/min)/(mg/L)]	Distance to Endpoint (miles)
130,000 - 140,000	4.8
140,000 - 160,000	5.0
160,000 - 180,000	5.2
180,000 - 190,000	5.4
190,000 - 210,000	5.6
210,000 - 220,000	5.8
220,000 - 240,000	6.0
240,000 - 261,000	6.2
261,000 - 325,000	6.8
325,000 - 397,000	7.5
397,000 - 477,000	8.1
477,000 - 566,000	8.7
566,000 - 663,000	9.3
663,000 - 769,000	9.9
769,000 - 1,010,000	11
1,010,000 - 1,280,000	12
1,280,000 - 1,600,000	14
1,600,000 - 1,950,000	15
1,950,000 - 2,340,000	16
2,340,000 - 2,770,000	17
2,770,000 - 3,240,000	19
3,240,000 - 4,590,000	22
4,590,000 - 6,190,000	25
>6,190,000	>25*

*Report distance as 25 miles

Reference Table 15
Neutrally Buoyant Plume Distances to Toxic Endpoint for Release Rate Divided by Endpoint
60-Minute Release, Rural Conditions, D Stability, Wind Speed 3.0 Meters per Second

Release Rate/Endpoint [(lbs/min)/(mg/L)]	Distance to Endpoint (miles)
0 - 79	0.1
79 - 630	0.2
630 - 1,600	0.3
1,600 - 2,800	0.4
2,800 - 5,200	0.6
5,200 - 7,900	0.8
7,900 - 11,000	1.0
11,000 - 14,000	1.2
14,000 - 19,000	1.4
19,000 - 23,000	1.6
23,000 - 27,000	1.8
27,000 - 32,000	2.0
32,000 - 36,000	2.2
36,000 - 42,000	2.4
42,000 - 47,000	2.6
47,000 - 52,000	2.8
52,000 - 57,000	3.0
57,000 - 61,000	3.2
61,000 - 68,000	3.4
68,000 - 73,000	3.6
73,000 - 79,000	3.8
79,000 - 84,000	4.0
84,000 - 91,000	4.2
91,000 - 97,000	4.4
97,000 - 100,000	4.6

Release Rate/Endpoint [(lbs/min)/(mg/L)]	Distance to Endpoint (miles)
100,000 - 108,000	4.8
108,000 - 113,000	5.0
113,000 - 120,000	5.2
120,000 - 126,000	5.4
126,000 - 132,000	5.6
132,000 - 140,000	5.8
140,000 - 150,000	6.0
150,000 - 151,000	6.2
151,000 - 171,000	6.8
171,000 - 191,000	7.5
191,000 - 212,000	8.1
212,000 - 233,000	8.7
233,000 - 256,000	9.3
256,000 - 280,000	9.9
280,000 - 332,000	11
332,000 - 390,000	12
390,000 - 456,000	14
456,000 - 529,000	15
529,000 - 610,000	16
610,000 - 699,000	17
699,000 - 796,000	19
796,000 - 1,080,000	22
1,080,000 - 1,410,000	25
>1,410,000	>25*

*Report distance as 25 miles

Reference Table 16
Neutrally Buoyant Plume Distances to Toxic Endpoint for Release Rate Divided by Endpoint
10-Minute Release, Urban Conditions, D Stability, Wind Speed 3.0 Meters per Second

Release Rate/Endpoint [(lbs/min)/(mg/L)]	Distance to Endpoint (miles)
0 - 160	0.1
160 - 1,400	0.2
1,400 - 3,600	0.3
3,600 - 6,900	0.4
6,900 - 13,000	0.6
13,000 - 22,000	0.8
22,000 - 31,000	1.0
31,000 - 42,000	1.2
42,000 - 59,000	1.4
59,000 - 73,000	1.6
73,000 - 88,000	1.8
88,000 - 100,000	2.0
100,000 - 120,000	2.2
120,000 - 150,000	2.4
150,000 - 170,000	2.6
170,000 - 200,000	2.8
200,000 - 230,000	3.0
230,000 - 260,000	3.2
260,000 - 310,000	3.4
310,000 - 340,000	3.6
340,000 - 390,000	3.8
390,000 - 430,000	4.0
430,000 - 490,000	4.2
490,000 - 540,000	4.4
540,000 - 600,000	4.6

Release Rate/Endpoint [(lbs/min)/(mg/L)]	Distance to Endpoint (miles)
600,000 - 660,000	4.8
660,000 - 720,000	5.0
720,000 - 810,000	5.2
810,000 - 880,000	5.4
880,000 - 950,000	5.6
950,000 - 1,000,000	5.8
1,000,000 - 1,100,000	6.0
1,100,000 - 1,220,000	6.2
1,220,000 - 1,530,000	6.8
1,530,000 - 1,880,000	7.5
1,880,000 - 2,280,000	8.1
2,280,000 - 2,710,000	8.7
2,710,000 - 3,200,000	9.3
3,200,000 - 3,730,000	9.9
3,730,000 - 4,920,000	11
4,920,000 - 6,310,000	12
6,310,000 - 7,890,000	14
7,890,000 - 9,660,000	15
9,660,000 - 11,600,000	16
11,600,000 - 13,800,000	17
13,800,000 - 16,200,000	19
16,200,000 - 23,100,000	22
23,100,000 - 31,300,000	25
>31,300,000	>25*

*Report distance as 25 miles

Reference Table 17
Neutrally Buoyant Plume Distances to Toxic Endpoint for Release Rate Divided by Endpoint
60-Minute Release, Urban Conditions, D Stability, Wind Speed 3.0 Meters per Second

Release Rate/Endpoint [(lbs/min)/(mg/L)]	Distance to Endpoint (miles)
0 - 200	0.1
200 - 1,700	0.2
1,700 - 4,500	0.3
4,500 - 8,600	0.4
8,600 - 17,000	0.6
17,000 - 27,000	0.8
27,000 - 39,000	1.0
39,000 - 53,000	1.2
53,000 - 73,000	1.4
73,000 - 90,000	1.6
90,000 - 110,000	1.8
110,000 - 130,000	2.0
130,000 - 150,000	2.2
150,000 - 170,000	2.4
170,000 - 200,000	2.6
200,000 - 220,000	2.8
220,000 - 240,000	3.0
240,000 - 270,000	3.2
270,000 - 300,000	3.4
300,000 - 320,000	3.6
320,000 - 350,000	3.8
350,000 - 370,000	4.0
370,000 - 410,000	4.2
410,000 - 430,000	4.4
430,000 - 460,000	4.6

Release Rate/Endpoint [(lbs/min)/(mg/L)]	Distance to Endpoint (miles)
460,000 - 490,000	4.8
490,000 - 520,000	5.0
520,000 - 550,000	5.2
550,000 - 580,000	5.4
580,000 - 610,000	5.6
610,000 - 640,000	5.8
640,000 - 680,000	6.0
680,000 - 705,000	6.2
705,000 - 804,000	6.8
804,000 - 905,000	7.5
905,000 - 1,010,000	8.1
1,010,000 - 1,120,000	8.7
1,120,000 - 1,230,000	9.3
1,230,000 - 1,350,000	9.9
1,350,000 - 1,620,000	11
1,620,000 - 1,920,000	12
1,920,000 - 2,250,000	14
2,250,000 - 2,620,000	15
2,620,000 - 3,030,000	16
3,030,000 - 3,490,000	17
3,490,000 - 3,980,000	19
3,980,000 - 5,410,000	22
5,410,000 - 7,120,000	25
>7,120,000	>25*

*Report distance as 25 miles

Reference Table 18
Dense Gas Distances to Toxic Endpoint
10-minute Release, Rural Conditions, D Stability, Wind Speed 3.0 Meters per Second

Release Rate (lbs/min)	Toxic Endpoint (mg/L)															
	0.0004	0.0007	0.001	0.002	0.0035	0.005	0.0075	0.01	0.02	0.035	0.05	0.075	0.1	0.25	0.5	0.75
	Distance (Miles)															
1	0.6	0.4	0.4	0.2	0.2	0.1	0.1	0.1	<0.1	<0.1	#	#	#	#	#	#
2	0.9	0.6	0.5	0.4	0.3	0.2	0.2	0.1	0.1	0.1	<0.1	<0.1	#	#	#	#
5	1.4	1.1	0.9	0.6	0.4	0.4	0.3	0.2	0.2	0.1	0.1	0.1	<0.1	#	#	#
10	2.0	1.5	1.2	0.9	0.6	0.5	0.4	0.4	0.2	0.2	0.1	0.1	0.1	<0.1	<0.1	#
30	3.7	2.7	2.2	1.5	1.1	0.9	0.7	0.7	0.5	0.3	0.3	0.2	0.2	0.1	0.1	<0.1
50	5.0	3.7	3.0	2.1	1.9	1.2	1.0	0.9	0.6	0.4	0.4	0.3	0.2	0.2	0.1	0.1
100	7.4	5.3	4.3	3.0	2.3	1.7	1.4	1.2	0.9	0.6	0.6	0.4	0.4	0.2	0.2	0.1
150	8.7	6.8	5.5	3.8	2.8	2.3	1.9	1.6	1.1	0.8	0.7	0.6	0.5	0.3	0.2	0.2
250	12	8.7	7.4	5.0	3.7	3.0	2.4	2.1	1.4	1.1	0.9	0.7	0.5	0.4	0.3	0.2
500	17	13	11	7.4	5.3	4.5	3.6	3.0	2.1	1.6	1.3	1.1	0.9	0.6	0.4	0.3
750	22	16	13	9.3	6.8	5.6	4.5	3.8	2.7	1.9	1.6	1.3	1.1	0.7	0.5	0.4
1,000	>25	19	16	11	8.1	6.8	5.2	4.5	3.1	2.3	2.2	1.5	1.3	0.8	0.6	0.4
1,500	*	23	19	13	9.9	8.1	6.8	5.6	3.9	2.9	2.4	1.9	1.6	1.0	0.7	0.6
2,000	*	>25	22	15	12	9.3	7.4	6.8	4.5	3.4	2.7	2.2	1.9	1.2	0.8	0.6
2,500	*	*	25	17	13	11	8.7	7.4	5.2	3.8	3.2	2.5	2.1	1.3	0.9	0.7
3,000	*	*	>25	19	14	12	9.3	8.1	5.7	4.2	3.5	2.8	2.4	1.4	1.0	0.8
4,000	*	*	*	22	17	14	11	9.3	6.8	4.9	4.1	3.3	2.8	1.7	1.1	0.9
5,000	*	*	*	>25	19	16	12	11	7.4	5.6	4.7	3.7	3.1	2.1	1.3	1.1
7,500	*	*	*	*	24	19	16	13	9.3	6.8	5.8	4.7	4.0	2.4	1.6	1.3
10,000	*	*	*	*	>25	22	18	16	11	8.1	6.8	5.3	4.6	2.8	1.9	1.5
15,000	*	*	*	*	*	>25	22	19	13	9.9	8.1	6.8	5.7	3.5	2.4	1.9
20,000	*	*	*	*	*	*	>25	22	16	11	9.3	7.4	6.8	4.0	2.8	2.2
50,000	*	*	*	*	*	*	*	>25	24	18	15	12	10	6.5	4.5	3.6
75,000	*	*	*	*	*	*	*	*	>25	22	18	15	13	7.8	5.4	4.4
100,000	*	*	*	*	*	*	*	*	*	>25	21	17	14	8.9	6.3	5.0
150,000	*	*	*	*	*	*	*	*	*	*	>25	20	17	11	7.4	6.0
200,000	*	*	*	*	*	*	*	*	*	*	*	23	19	12	8.5	6.8

* >25 miles (report distance as 25 miles)

\# <0.1 mile (report distance as 0.1 mile)

Reference Table 19
Dense Gas Distances to Toxic Endpoint
60-minute Release, Rural Conditions, D Stability, Wind Speed 3.0 Meters per Second

Release Rate (lbs/min)	Toxic Endpoint (mg/L)															
	0.0004	0.0007	0.001	0.002	0.0035	0.005	0.0075	0.01	0.02	0.035	0.05	0.075	0.1	0.25	0.5	0.75
	Distance (Miles)															
1	0.5	0.4	0.3	0.2	0.2	0.1	0.1	0.1	<0.1	#	#	#	#	#	#	#
2	0.8	0.6	0.5	0.3	0.2	0.2	0.2	0.1	0.1	<0.1	<0.1	<0.1	#	#	#	#
5	1.6	1.0	0.8	0.5	0.4	0.3	0.2	0.2	0.2	0.1	0.1	0.1	<0.1	#	#	#
10	2.0	1.4	1.2	0.8	0.6	0.5	0.4	0.3	0.2	0.2	0.1	0.1	0.1	<0.1	<0.1	#
30	4.0	2.8	2.2	1.5	1.1	0.9	0.7	0.6	0.4	0.3	0.2	0.2	0.2	0.1	0.1	<0.1
50	5.5	3.9	3.1	2.1	1.5	1.2	1.0	0.8	0.6	0.4	0.3	0.3	0.2	0.1	0.1	0.1
100	8.7	6.1	4.8	3.2	2.2	1.8	1.4	1.2	0.8	0.6	0.5	0.4	0.3	0.2	0.1	0.1
150	12	8.1	6.2	4.1	2.9	2.3	1.8	1.6	1.1	0.7	0.6	0.5	0.4	0.3	0.2	0.1
250	17	11	8.7	5.6	4.0	3.2	2.5	2.1	1.4	1.1	0.9	0.7	0.6	0.4	0.2	0.2
500	>25	19	14	9.3	6.2	5.0	3.9	3.3	2.2	1.6	1.3	1.0	0.9	0.5	0.4	0.3
750	*	25	19	12	8.7	6.8	5.1	4.2	2.8	2.0	1.6	1.3	1.1	0.6	0.4	0.4
1,000	*	>25	24	15	11	8.1	6.1	5.2	3.4	2.4	1.9	1.5	1.3	0.7	0.5	0.4
1,500	*	*	>25	20	14	11	8.1	6.8	4.3	3.0	2.5	1.9	1.7	1.0	0.7	0.5
2,000	*	*	*	24	17	13	9.9	8.1	5.2	3.7	2.9	2.3	1.9	1.2	0.7	0.6
2,500	*	*	*	>25	19	15	12	9.3	6.0	4.3	3.4	2.7	2.2	1.3	0.9	0.7
3,000	*	*	*	*	22	17	13	11	6.8	4.8	3.8	3.0	2.5	1.5	1.0	0.8
4,000	*	*	*	*	>25	21	16	14	8.7	5.8	4.7	3.6	3.0	1.7	1.2	0.9
5,000	*	*	*	*	*	25	19	16	9.9	6.8	5.3	4.1	3.5	2.0	1.4	1.1
7,500	*	*	*	*	*	>25	25	20	13	9.3	6.8	5.4	4.5	2.6	1.7	1.4
10,000	*	*	*	*	*	*	>25	25	16	11	8.7	6.8	5.4	3.1	2.1	1.6
15,000	*	*	*	*	*	*	*	>25	21	14	11	8.7	7.4	4.0	2.6	2.1
20,000	*	*	*	*	*	*	*	*	25	17	14	11	8.7	4.8	3.1	2.5
50,000	*	*	*	*	*	*	*	*	>25	>25	25	19	16	8.8	5.6	4.3
75,000	*	*	*	*	*	*	*	*	*	*	>25	25	20	11	7.3	5.6
100,000	*	*	*	*	*	*	*	*	*	*	*	>25	24	14	9.4	6.8
150,000	*	*	*	*	*	*	*	*	*	*	*	*	>25	17	11	8.7
200,000	*	*	*	*	*	*	*	*	*	*	*	*	*	20	13	10

* >25 miles (report distance as 25 miles)

\# <0.1 mile (report distance as 0.1 mile)

Reference Table 20
Dense Gas Distances to Toxic Endpoint
10-minute Release, Urban Conditions, D Stability, Wind Speed 3.0 Meters per Second

Release Rate (lbs/min)	Toxic Endpoint (mg/L)															
	0.0004	0.0007	0.001	0.002	0.0035	0.005	0.0075	0.01	0.02	0.035	0.05	0.075	0.1	0.25	0.5	0.75
	Distance (Miles)															
1	0.5	0.3	0.2	0.2	0.1	0.1	0.1	0.1	<0.1	#	#	#	#	#	#	#
2	0.7	0.5	0.4	0.3	0.2	0.2	0.1	0.1	0.1	<0.1	<0.1	#	#	#	#	#
5	1.1	0.8	0.6	0.5	0.3	0.3	0.2	0.2	0.1	0.1	0.1	<0.1	<0.1	#	#	#
10	2.1	1.2	1.0	0.7	0.5	0.4	0.3	0.3	0.2	0.1	0.1	0.1	0.1	<0.1	#	#
30	3.0	2.2	1.9	1.2	0.9	0.8	0.6	0.6	0.4	0.3	0.2	0.2	0.1	0.1	<0.1	#
50	4.1	3.0	2.5	1.6	1.2	1.0	0.8	0.7	0.5	0.3	0.3	0.2	0.2	0.1	0.1	<0.1
100	5.8	4.3	3.5	2.7	1.8	1.4	1.2	1.0	0.7	0.6	0.4	0.4	0.3	0.2	0.1	0.1
150	7.4	5.5	4.5	3.1	2.2	1.9	1.4	1.2	0.9	0.7	0.6	0.4	0.4	0.2	0.2	0.1
250	9.9	7.4	5.8	4.1	3.0	2.5	2.0	1.7	1.1	0.9	0.7	0.6	0.5	0.3	0.2	0.1
500	14	11	8.7	5.9	4.3	3.6	2.9	2.5	1.7	1.2	1.0	0.8	0.7	0.4	0.3	0.2
750	17	13	11	7.4	5.5	4.5	3.6	3.1	2.1	1.6	1.2	1.0	0.9	0.5	0.4	0.3
1,000	20	15	12	8.7	6.2	5.3	4.3	3.5	2.5	1.8	1.5	1.2	1.0	0.6	0.4	0.3
1,500	>25	19	16	11	8.1	6.2	5.2	4.5	3.0	2.2	1.8	1.5	1.2	0.7	0.5	0.4
2,000	*	22	18	12	9.3	7.4	6.2	5.2	3.7	2.7	2.2	1.7	1.4	0.9	0.6	0.5
2,500	*	24	20	14	11	8.7	6.8	6.0	3.8	3.0	2.2	1.9	1.7	1.0	0.7	0.6
3,000	*	>25	22	16	11	9.3	7.4	6.8	4.5	3.3	2.7	2.1	1.9	1.1	0.7	0.6
4,000	*	*	>25	18	14	11	8.7	7.4	5.3	4.0	3.2	2.6	2.1	1.2	0.9	0.7
5,000	*	*	*	20	15	12	9.9	8.7	5.8	4.4	3.6	2.9	2.4	1.4	0.9	0.7
7,500	*	*	*	>25	19	16	12	11	7.4	5.5	4.5	3.6	3.0	1.8	1.2	0.9
10,000	*	*	*	*	22	18	14	12	8.7	6.2	5.2	4.2	3.6	2.1	1.4	1.1
15,000	*	*	*	*	>25	22	18	16	11	8.1	6.8	5.2	4.4	2.6	1.7	1.3
20,000	*	*	*	*	*	>25	20	18	12	9.3	7.4	6.0	5.2	3.0	2.0	1.6
50,000	*	*	*	*	*	*	>25	>25	20	15	12	9.7	8.3	5.0	3.3	2.6
75,000	*	*	*	*	*	*	*	*	25	18	15	12	10	6.1	4.1	3.1
100,000	*	*	*	*	*	*	*	*	>25	21	17	14	12	7.0	4.7	3.7
150,000	*	*	*	*	*	*	*	*	*	>25	21	17	14	8.5	5.7	4.5
200,000	*	*	*	*	*	*	*	*	*	*	24	19	16	9.7	6.5	5.1

* >25 miles (report distance as 25 miles)

\# <0.1 mile (report distance as 0.1 mile)

Reference Table 21
Dense Gas Distances to Toxic Endpoint
60-minute Release, Urban Conditions, D Stability, Wind Speed 3.0 Meters per Second

Release Rate (lbs/min)	Toxic Endpoint (mg/L)															
	0.0004	0.0007	0.001	0.002	0.0035	0.005	0.0075	0.01	0.02	0.035	0.05	0.075	0.1	0.25	0.5	0.75
	Distance (Miles)															
1	0.4	0.3	0.2	0.2	0.1	0.1	0.1	<0.1	#	#	#	#	#	#	#	#
2	0.7	0.5	0.4	0.2	0.2	0.2	0.1	0.1	<0.1	<0.1	#	#	#	#	#	#
5	1.1	0.8	0.7	0.4	0.3	0.2	0.2	0.2	0.1	0.1	<0.1	<0.1	<0.1	#	#	#
10	1.7	1.2	1.0	0.7	0.5	0.4	0.3	0.3	0.2	0.1	0.1	0.1	0.1	<0.1	#	#
30	3.3	2.4	1.9	1.3	0.9	0.7	0.6	0.5	0.3	0.2	0.2	0.2	0.1	0.1	<0.1	#
50	4.7	3.3	2.6	1.7	1.2	1.0	0.8	0.7	0.4	0.3	0.3	0.2	0.2	0.1	0.1	<0.1
100	7.4	5.2	4.1	2.7	1.9	1.5	1.2	1.0	0.7	0.5	0.4	0.3	0.3	0.2	0.1	0.1
150	9.9	6.8	5.3	3.4	2.4	1.9	1.5	1.3	0.9	0.6	0.5	0.4	0.3	0.2	0.1	0.1
250	14	9.3	7.4	4.7	3.4	2.7	2.1	1.7	1.1	0.8	0.7	0.5	0.4	0.3	0.2	0.1
500	22	16	12	7.4	5.2	4.2	3.2	2.7	1.7	1.2	1.0	0.8	0.7	0.4	0.2	0.2
750	>25	20	16	9.9	6.8	5.4	4.2	3.5	2.2	1.6	1.3	1.0	0.9	0.5	0.3	0.3
1,000	*	24	19	12	8.1	6.8	5.0	4.2	2.7	1.8	1.6	1.2	1.0	0.6	0.4	0.3
1,500	*	>25	>25	16	11	8.7	6.8	5.5	3.5	1.9	2.0	1.6	1.3	0.7	0.5	0.4
2,000	*	*	*	19	14	11	8.1	6.8	4.2	3.0	2.2	1.9	1.6	0.9	0.6	0.4
2,500	*	*	*	23	16	12	9.3	7.4	4.9	3.4	2.7	2.1	1.7	1.0	0.6	0.5
3,000	*	*	*	>25	18	14	11	8.7	5.5	3.8	3.0	2.4	2.0	1.1	0.7	0.6
4,000	*	*	*	*	22	17	13	11	6.8	4.7	3.1	2.8	2.4	1.3	0.9	0.7
5,000	*	*	*	*	>25	20	16	12	8.1	5.3	4.3	3.3	2.7	1.5	1.0	0.7
7,500	*	*	*	*	*	25	20	17	11	6.8	5.6	4.3	3.5	2.0	1.2	0.9
10,000	*	*	*	*	*	>25	24	20	13	8.7	6.8	5.2	4.3	2.4	1.5	1.1
15,000	*	*	*	*	*	*	>25	>25	17	11	8.7	6.8	5.6	3.0	1.9	1.5
20,000	*	*	*	*	*	*	*	*	20	14	11	8.1	6.8	3.6	2.3	1.7
50,000	*	*	*	*	*	*	*	*	>25	>25	20	15	13	6.6	4.0	3.1
75,000	*	*	*	*	*	*	*	*	*	*	>25	20	16	8.7	5.3	3.9
100,000	*	*	*	*	*	*	*	*	*	*	*	24	20	10	6.3	4.7
150,000	*	*	*	*	*	*	*	*	*	*	*	>25	>25	14	8.2	6.1
200,000	*	*	*	*	*	*	*	*	*	*	*	*	*	16	9.9	7.3

* >25 miles (report distance as 25 miles)

\# <0.1 mile (report distance as 0.1 mile)

Reference Table 22
Distances to Toxic Endpoint for Anhydrous Ammonia Liquefied Under Pressure
D Stability, Wind Speed 3.0 Meters per Second

Release Rate (lbs/min)	Distance to Endpoint (miles)	
	Rural	Urban
<10	<0.1*	<0.1*
10	0.1	
15	0.1	
20	0.1	
30	0.1	
40	0.1	
50	0.1	
60	0.2	0.1
70	0.2	0.1
80	0.2	0.1
90	0.2	0.1
100	0.2	0.1
150	0.2	0.1
200	0.3	0.1
250	0.3	0.1
300	0.3	0.1
400	0.4	0.2
500	0.4	0.2
600	0.5	0.2
700	0.5	0.2
750	0.5	0.2
800	0.5	0.2

Release Rate (lbs/min)	Distance to Endpoint (miles)	
	Rural	Urban
900	0.6	0.2
1,000	0.6	0.2
1,500	0.7	0.3
2,000	0.8	0.3
2,500	0.9	0.3
3,000	1.0	0.4
4,000	1.2	0.4
5,000	1.3	0.5
7,500	1.6	0.5
10,000	1.8	0.6
15,000	2.2	0.7
20,000	2.5	0.8
25,000	2.8	0.9
30,000	3.1	1.0
40,000	3.5	1.1
50,000	3.9	1.2
75,000	4.8	1.4
100,000	5.4	1.6
150,000	6.6	1.9
200,000	7.6	2.1
250,000	8.4	2.3

* Report distance as 0.1 mile

Reference Table 23
Distances to Toxic Endpoint for Non-liquefied Ammonia, Ammonia Liquefied by Refrigeration, or Aqueous Ammonia
D Stability, Wind Speed 3.0 Meters per Second

Release Rate (lbs/min)	Distance to Endpoint (miles)	
	Rural	Urban
<8	<0.1*	<0.1*
8	0.1	
10	0.1	
15	0.1	
20	0.1	
30	0.1	
40	0.1	
50	0.2	0.1
60	0.2	0.1
70	0.2	0.1
80	0.2	0.1
90	0.2	0.1
100	0.2	0.1
150	0.3	0.1
200	0.3	0.1
250	0.4	0.2
300	0.4	0.2
400	0.4	0.2
500	0.5	0.2
600	0.6	0.2
700	0.6	0.2
750	0.6	0.2
800	0.7	0.2
900	0.7	0.3
1,000	0.8	0.3
1,500	1.0	0.4
2,000	1.2	0.4
2,500	1.2	0.4
3,000	1.5	0.5
4,000	1.8	0.6
5,000	2.0	0.7
7,500	2.2	0.7
10,000	2.5	0.8
15,000	3.1	1.0
20,000	3.6	1.2
25,000	4.1	1.3
30,000	4.4	1.4
40,000	5.1	1.6
50,000	5.8	1.8
75,000	7.1	2.2
100,000	8.2	2.5
150,000	10	3.1
200,000	12	3.5

* Report distance as 0.1 mile

Reference Table 24
Distances to Toxic Endpoint for Chlorine
D Stability, Wind Speed 3.0 Meters per Second

Release Rate (lbs/min)	Distance to Endpoint (miles)	
	Rural	Urban
1	<0.1*	<0.1*
2	0.1	
5	0.1	
10	0.2	0.1
15	0.2	0.1
20	0.2	0.1
30	0.3	0.1
40	0.3	0.1
50	0.3	0.1
60	0.4	0.2
70	0.4	0.2
80	0.4	0.2
90	0.4	0.2
100	0.5	0.2
150	0.6	0.2
200	0.6	0.3
250	0.7	0.3
300	0.8	0.3
400	0.8	0.4
500	1.0	0.4
600	1.0	0.4
700	1.1	0.4

Release Rate (lbs/min)	Distance to Endpoint (miles)	
	Rural	Urban
750	1.2	0.4
800	1.2	0.5
900	1.2	0.5
1,000	1.3	0.5
1,500	1.6	0.6
2,000	1.8	0.6
2,500	2.0	0.7
3,000	2.2	0.8
4,000	2.5	0.8
5,000	2.8	0.9
7,500	3.4	1.2
10,000	3.9	1.3
15,000	4.6	1.6
20,000	5.3	1.8
25,000	5.9	2.0
30,000	6.4	2.1
40,000	7.3	2.4
50,000	8.1	2.7
75,000	9.8	3.2
100,000	11	3.6
150,000	13	4.2
200,000	15	4.8

* Report distance as 0.1 mile

Reference Table 25
Distances to Toxic Endpoint for Sulfur Dioxide
D Stability, Wind Speed 3.0 Meters per Second

Release Rate (lbs/min)	Distance to Endpoint (miles)	
	Rural	Urban
1	<0.1*	<0.1*
2	0.1	
5	0.1	
10	0.2	0.1
15	0.2	0.1
20	0.2	0.1
30	0.2	0.1
40	0.3	0.1
50	0.3	0.1
60	0.4	0.2
70	0.4	0.2
80	0.4	0.2
90	0.4	0.2
100	0.5	0.2
150	0.6	0.2
200	0.6	0.2
250	0.7	0.3
300	0.8	0.3
400	0.9	0.4
500	1.0	0.4
600	1.1	0.4
700	1.2	0.4

Release Rate (lbs/min)	Distance to Endpoint (miles)	
	Rural	Urban
750	1.3	0.5
800	1.3	0.5
900	1.4	0.5
1,000	1.5	0.5
1,500	1.9	0.6
2,000	2.2	0.7
2,500	2.3	0.8
3,000	2.7	0.8
4,000	3.1	1.0
5,000	3.3	1.1
7,500	4.0	1.3
10,000	4.6	1.4
15,000	5.6	1.7
20,000	6.5	1.9
25,000	7.3	2.1
30,000	8.0	2.3
40,000	9.2	2.6
50,000	10	2.9
75,000	13	3.5
100,000	14	4.0
150,000	18	4.7
200,000	20	5.4

* Report distance as 0.1 mile

Reference Table 26
Neutrally Buoyant Plume Distances to Lower Flammability Limit (LFL) For Release Rate Divided by LFL Rural Conditions, D Stability, Wind Speed 3.0 Meters per Second

Release Rate/Endpoint [(lbs/min)/(mg/L)]	Distance to Endpoint (miles)
0 - 28	0.1
28 - 40	0.1
40 - 60	0.1
60 - 220	0.2
220 - 530	0.3
530 - 860	0.4
860 - 1,300	0.5
1,300 - 1,700	0.6
1,700 - 2,200	0.7
2,200 - 2,700	0.8

Release Rate/Endpoint [(lbs/min)/(mg/L)]	Distance to Endpoint (miles)
2,700 - 3,300	0.9
3,300 - 3,900	1.0
3,900 - 4,500	1.1
4,500 - 5,200	1.2
5,200 - 5,800	1.3
5,800 - 6,800	1.4
6,800 - 8,200	1.6
8,200 - 9,700	1.8
9,700 - 11,000	2.0
11,000 - 13,000	2.2

Reference Table 27
Neutrally Buoyant Plume Distances to Lower Flammability Limit (LFL) For Release Rate Divided by LFL Urban Conditions, D Stability, Wind Speed 3.0 Meters per Second

Release Rate/Endpoint [(lbs/min)/(mg/L)]	Distance to Endpoint (miles)
0 - 68	0.1
68 - 100	0.1
100 - 150	0.1
150 - 710	0.2
710 - 1,500	0.3
1,500 - 2,600	0.4
2,600 - 4,000	0.5
4,000 - 5,500	0.6

Release Rate/Endpoint [(lbs/min)/(mg/L)]	Distance to Endpoint (miles)
5,500 - 7,300	0.7
7,300 - 9,200	0.8
9,200 - 11,000	0.9
11,000 - 14,000	1.0
14,000 - 18,000	1.2
18,000 - 26,000	1.4
26,000 - 31,000	1.6
31,000 - 38,000	1.8

Reference Table 28
Dense Gas Distances to Lower Flammability Limit
Rural Conditions, D Stability, Wind Speed 3.0 Meters per Second

Release Rate (lbs/min)	Lower Flammability Limit (mg/L)									
	27	30	35	40	45	50	60	70	100	>100
	Distance (Miles)									
<1,500	#	#	#	#	#	#	#	#	#	#
1,500	<0.1	<0.1	#	#	#	#	#	#	#	#
2,000	0.1	0.1	<0.1	#	#	#	#	#	#	#
2,500	0.1	0.1	0.1	<0.1	#	#	#	#	#	#
3,000	0.1	0.1	0.1	0.1	<0.1	<0.1	#	#	#	#
4,000	0.1	0.1	0.1	0.1	0.1	0.1	<0.1	#	#	#
5,000	0.1	0.1	0.1	0.1	0.1	0.1	0.1	<0.1	#	#
7,500	0.2	0.1	0.1	0.1	0.1	0.1	0.1	0.1	<0.1	#
10,000	0.2	0.2	0.1	0.1	0.1	0.1	0.1	0.1	0.1	<0.1

< 0.1 mile (report distance as 0.1 mile)

Reference Table 29
Dense Gas Distances to Lower Flammability Limit
Urban Conditions, D Stability, Wind Speed 3.0 Meters per Second

Release Rate (lbs/min)	Lower Flammability Limit (mg/L)				
	27	30	35	40	>40
	Distance (Miles)				
<5,000	#	#	#	#	#
5,000	<0.1	<0.1	#	#	#
7,500	0.1	0.1	<0.1	#	#
10,000	0.1	0.1	0.1	<0.1	#

< 0.1 mile (report distance as 0.1 mile)

Reference Table 30
Distance to Radiant Heat Dose at Potential Second Degree Burn Threshold Assuming Exposure for Duration of Fireball from BLEVE
(Dose = $[5\ kW/m^2]^{4/3}$ x Exposure Time)

Quantity in Fireball (pounds)		1,000	5,000	10,000	20,000	30,000	50,000	75,000	100,000	200,000	300,000	500,000
Duration of Fireball (seconds)		3.5	5.9	7.5	9.4	10.8	12.7	14.8	15.5	17.4	18.7	20.3
CAS No.	Chemical Name	Distance (miles) at which Exposure for Duration of Fireball May Cause Second Degree Burns										
75-07-0	Acetaldehyde	0.04	0.08	0.1	0.1	0.2	0.2	0.3	0.3	0.4	0.5	0.6
74-86-2	Acetylene	0.05	0.1	0.1	0.2	0.2	0.3	0.4	0.4	0.5	0.6	0.8
598-73-2	Bromotrifluoroethylene	0.01	0.02	0.03	0.04	0.05	0.06	0.07	0.08	0.1	0.1	0.2
106-99-0	1,3-Butadiene	0.05	0.1	0.1	0.2	0.2	0.3	0.4	0.4	0.5	0.6	0.8
106-97-8	Butane	0.05	0.1	0.1	0.2	0.2	0.3	0.4	0.4	0.5	0.6	0.8
106-98-9	1-Butene	0.05	0.1	0.1	0.2	0.2	0.3	0.4	0.4	0.5	0.6	0.8
107-01-7	2-Butene	0.05	0.1	0.1	0.2	0.2	0.3	0.4	0.4	0.5	0.6	0.8
25167-67-3	Butene	0.05	0.1	0.1	0.2	0.2	0.3	0.4	0.4	0.5	0.6	0.8
590-18-1	2-Butene-cis	0.05	0.1	0.1	0.2	0.2	0.3	0.4	0.4	0.5	0.6	0.8
624-64-6	2-Butene-trans	0.05	0.1	0.1	0.2	0.2	0.3	0.4	0.4	0.5	0.6	0.8
463-58-1	Carbon oxysulfide	0.02	0.05	0.06	0.09	0.1	0.1	0.2	0.2	0.2	0.3	0.3
7791-21-1	Chlorine monoxide	0.01	0.02	0.02	0.03	0.03	0.04	0.05	0.06	0.08	0.09	0.1
557-98-2	2-Chloropropylene	0.03	0.07	0.1	0.1	0.2	0.2	0.2	0.3	0.4	0.4	0.5
590-21-6	1-Chloropropylene	0.03	0.07	0.1	0.1	0.2	0.2	0.2	0.3	0.4	0.4	0.5
460-19-5	Cyanogen	0.03	0.07	0.1	0.1	0.2	0.2	0.2	0.3	0.4	0.4	0.5
75-19-4	Cyclopropane	0.05	0.1	0.1	0.2	0.2	0.3	0.4	0.4	0.5	0.6	0.8
4109-96-0	Dichlorosilane	0.02	0.04	0.06	0.08	0.1	0.1	0.2	0.2	0.2	0.3	0.3
75-37-6	Difluoroethane	0.02	0.05	0.07	0.1	0.1	0.1	0.2	0.2	0.3	0.3	0.4
124-40-3	Dimethylamine	0.04	0.09	0.1	0.2	0.2	0.3	0.3	0.4	0.5	0.5	0.7
463-82-1	2,2-Dimethylpropane	0.05	0.1	0.1	0.2	0.2	0.3	0.4	0.4	0.5	0.6	0.8
74-84-0	Ethane	0.05	0.1	0.1	0.2	0.2	0.3	0.4	0.4	0.5	0.6	0.8
107-00-6	Ethyl acetylene	0.05	0.1	0.1	0.2	0.2	0.3	0.4	0.4	0.5	0.6	0.8

Reference Table 30 (continued)

Quantity in Fireball (pounds)		1,000	5,000	10,000	20,000	30,000	50,000	75,000	100,000	200,000	300,000	500,000
Duration of Fireball (seconds)		3.5	5.9	7.5	9.4	10.8	12.7	14.8	15.5	17.4	18.7	20.3
CAS No.	Chemical Name	Distance (miles) at which Exposure for Duration of Fireball May Cause Second Degree Burns										
75-04-7	Ethylamine	0.04	0.09	0.1	0.2	0.2	0.3	0.3	0.4	0.5	0.5	0.7
75-00-3	Ethyl chloride	0.03	0.07	0.09	0.1	0.2	0.2	0.2	0.3	0.3	0.4	0.5
74-85-1	Ethylene	0.05	0.1	0.1	0.2	0.2	0.3	0.4	0.4	0.5	0.6	0.8
60-29-7	Ethyl ether	0.04	0.09	0.1	0.2	0.2	0.2	0.3	0.3	0.5	0.5	0.7
75-08-1	Ethyl mercaptan	0.04	0.08	0.1	0.2	0.2	0.2	0.3	0.3	0.4	0.5	0.6
109-95-5	Ethyl nitrite	0.03	0.06	0.09	0.1	0.1	0.2	0.2	0.3	0.3	0.4	0.5
1333-74-0	Hydrogen	0.08	0.2	0.2	0.3	0.4	0.5	0.6	0.6	0.9	1.0	1.2
75-28-5	Isobutane	0.05	0.1	0.1	0.2	0.2	0.3	0.4	0.4	0.5	0.6	0.8
78-78-4	Isopentane	0.05	0.1	0.1	0.2	0.2	0.3	0.4	0.4	0.5	0.6	0.8
78-79-5	Isoprene	0.05	0.1	0.1	0.2	0.2	0.3	0.4	0.4	0.5	0.6	0.7
75-31-0	Isopropylamine	0.04	0.09	0.1	0.2	0.2	0.3	0.3	0.4	0.5	0.6	0.7
75-29-6	Isopropyl chloride	0.04	0.07	0.1	0.1	0.2	0.2	0.3	0.3	0.4	0.4	0.5
74-82-8	Methane	0.05	0.1	0.1	0.2	0.2	0.3	0.4	0.4	0.6	0.6	0.8
74-89-5	Methylamine	0.04	0.08	0.1	0.2	0.2	0.2	0.3	0.3	0.4	0.5	0.6
563-45-1	3-Methyl-1-butene	0.05	0.1	0.1	0.2	0.2	0.3	0.4	0.4	0.5	0.6	0.8
563-46-2	2-Methyl-1-butene	0.05	0.1	0.1	0.2	0.2	0.3	0.4	0.4	0.5	0.6	0.7
115-10-6	Methyl ether	0.04	0.08	0.1	0.2	0.2	0.2	0.3	0.3	0.4	0.5	0.6
107-31-3	Methyl formate	0.03	0.06	0.08	0.1	0.1	0.2	0.2	0.2	0.3	0.4	0.4
115-11-7	2-Methylpropene	0.05	0.1	0.1	0.2	0.2	0.3	0.4	0.4	0.5	0.6	0.8
504-60-9	1,3-Pentadiene	0.05	0.1	0.1	0.2	0.2	0.3	0.3	0.4	0.5	0.6	0.7
109-66-0	Pentane	0.05	0.1	0.1	0.2	0.2	0.3	0.4	0.4	0.5	0.6	0.8
109-67-1	1-Pentene	0.05	0.1	0.1	0.2	0.2	0.3	0.4	0.4	0.5	0.6	0.8
646-04-8	2-Pentene, (E)-	0.05	0.1	0.1	0.2	0.2	0.3	0.4	0.4	0.5	0.6	0.8

Reference Table 30 (continued)

Quantity in Fireball (pounds)		1,000	5,000	10,000	20,000	30,000	50,000	75,000	100,000	200,000	300,000	500,000
Duration of Fireball (seconds)		3.5	5.9	7.5	9.4	10.8	12.7	14.8	15.5	17.4	18.7	20.3
CAS No.	Chemical Name	Distance (miles) at which Exposure for Duration of Fireball May Cause Second Degree Burns										
627-20-3	2-Pentene, (Z)-	0.05	0.1	0.1	0.2	0.2	0.3	0.4	0.4	0.5	0.6	0.8
463-49-0	Propadiene	0.05	0.1	0.1	0.2	0.2	0.3	0.4	0.4	0.5	0.6	0.8
74-98-6	Propane	0.05	0.1	0.1	0.2	0.2	0.3	0.4	0.4	0.5	0.6	0.8
115-07-1	Propylene	0.05	0.1	0.1	0.2	0.2	0.3	0.4	0.4	0.5	0.6	0.8
74-99-7	Propyne	0.05	0.1	0.1	0.2	0.2	0.3	0.4	0.4	0.5	0.6	0.8
7803-62-5	Silane	0.05	0.1	0.1	0.2	0.2	0.3	0.4	0.4	0.5	0.6	0.7
116-14-3	Tetrafluoroethylene	0.01	0.02	0.02	0.03	0.04	0.05	0.06	0.07	0.09	0.1	0.1
75-76-3	Tetramethylsilane	0.05	0.1	0.1	0.2	0.2	0.3	0.3	0.4	0.5	0.6	0.7
10025-78-2	Trichlorosilane	0.01	0.03	0.04	0.06	0.07	0.08	0.1	0.1	0.2	0.2	0.2
79-38-9	Trifluorochloroethylene	0.01	0.02	0.03	0.04	0.05	0.06	0.07	0.08	0.1	0.1	0.2
75-50-3	Trimethylamine	0.04	0.09	0.1	0.2	0.2	0.3	0.3	0.4	0.5	0.6	0.7
689-97-4	Vinyl acetylene	0.05	0.1	0.1	0.2	0.2	0.3	0.4	0.4	0.5	0.6	0.8
75-01-4	Vinyl chloride	0.03	0.07	0.09	0.1	0.2	0.2	0.2	0.3	0.3	0.4	0.5
109-92-2	Vinyl ethyl ether	0.04	0.09	0.1	0.2	0.2	0.2	0.3	0.3	0.4	0.5	0.6
75-02-5	Vinyl fluoride	0.01	0.02	0.03	0.04	0.05	0.06	0.08	0.09	0.1	0.1	0.2
75-35-4	Vinylidene chloride	0.02	0.05	0.07	0.09	0.1	0.1	0.2	0.2	0.3	0.3	0.4
75-38-7	Vinylidene fluoride	0.02	0.05	0.07	0.09	0.1	0.1	0.2	0.2	0.3	0.3	0.4
107-25-5	Vinyl methyl ether	0.04	0.08	0.1	0.2	0.2	0.2	0.3	0.3	0.4	0.5	0.6

11 ESTIMATING OFFSITE RECEPTORS

In Chapter 11

- How to estimate the number of offsite receptors potentially affected by your worst-case and alternative scenarios.
- Where to find the data you need.

The rule requires that you estimate residential populations within the circle defined by the endpoint for your worst-case and alternative release scenarios. In addition, you must report in the RMP whether certain types of public receptors and environmental receptors are within the circles.

To estimate residential populations, you may use the most recent Census data or any other source of data that you believe is more accurate. Local authorities may be able to provide information on offsite receptors. You are not required to update Census data or conduct any surveys to develop your estimates. Census data are available in public libraries and in the LandView system, which is available on CD-ROM (see box below). The rule requires that you estimate populations to two significant digits. For example, if there are 1,260 people within the circle, you may report 1,300 people. If the number of people is between 10 and 100, estimate to the nearest 10. If the number of people is less than 10, provide the actual number.

How to obtain Census data and LandView

Census data can be found in publications of the Bureau of the Census, available in public libraries, including *County and City Data Book*.

LandView ®III is a desktop mapping system that includes database extracts from EPA, the Bureau of the Census, the U.S. Geological Survey, the Nuclear Regulatory Commission, the Department of Transportation, and the Federal Emergency Management Agency. These databases are presented in a geographic context on maps that show jurisdictional boundaries, detailed networks of roads, rivers, and railroads, census block group and tract polygons, schools, hospitals, churches, cemeteries, airports, dams, and other landmark features.

CD-ROM for IBM-compatible PCS
CD-TGR95-LV3-KIT $99 per disc (by region) or $549 for 11 disc set

U.S. Department of Commerce
Bureau of the Census
P.O. Box 277943
Atlanta, GA 30384-7943
Phone: 301-457-4100 (Customer Services -- orders)
Fax: (888) 249-7295 (toll-free)
Fax: (301) 457-3842 (local)
Phone: (301) 457-1128 (Geography Staff -- content)
http://www.census.gov/ftp/pub/geo/www/tiger/

Further information on LandView and other sources of Census data is available at the Bureau of the Census web site at www.census.gov.

Census data are presented by Census tract. If your circle covers only a portion of the tract, you should develop an estimate for that portion. The easiest way to do this is to determine the population density per square mile (total population of the Census tract divided by the number of square miles in the tract) and apply that density figure to the number of square miles within your circle. Because there is likely to be considerable variation in actual densities within a Census tract, this number will be approximate. The rule, however, does not require you to correct the number.

Other public receptors must be noted in the RMP. If there are any schools, residences, hospitals, prisons, public recreational areas, or commercial, office, or industrial areas within the circle, you must report that. Any of these locations inhabited or occupied by the public at any time without restriction by the source is a public receptor. You are not required to develop a list of all institutions and areas; you must simply check off which types of receptors are within the circle. Most of these institutions or areas can be identified from local street maps. Recreational areas include public swimming pools, public parks, and other areas that are used for recreational activities (e.g., baseball fields). Commercial and industrial areas include shopping malls, strip malls, downtown business areas, industrial parks, etc. See EPA's *General Guidance for Risk Management Programs (40 CFR part 68)* for further information on identifying public receptors.

Environmental receptors are defined as national or state parks, forests, or monuments; officially designated wildlife sanctuaries, preserves, or refuges; and Federal wilderness areas. All of these can be identified on local U.S. Geological Survey (USGS) maps (see box below). You are not required to locate each of these specifically. You are only required to check off in the RMP that these specific types of areas are within the circle. If any part of one of these receptors is within your circles, you must note that in the RMP.

Important: The rule does not require you to assess the likelihood, type, or severity of potential impacts on either public or environmental receptors. Identifying them as within the circle simply indicates that they could be adversely affected by the release.

How to obtain USGS maps

The production of digital cartographic data and graphic maps comprises the largest component of the USGS National Mapping Program. The USGS's most familiar product is the 1:24,000-scale Topographic Quadrangle Map. This is the primary scale of data produced, and depicts greater detail for a smaller area than intermediate-scale (1:50,000 and 1:100,000) and small-scale (1:250,000, 1:2,000,000 or smaller) products, which show selectively less detail for larger areas.

U.S. Geological Survey
508 National Center
12201 Sunrise Valley Drive
Reston, VA 20192
Phone: (703) 648-4000
http://mapping.usgs.gov

To order USGS maps by fax, select, print, and complete one of the online forms and fax to 303-202-4693.

A list of the nearest commercial dealers is available at: http://mapping.usgs.gov/esic/usimage/dealers.html

For more information or ordering assistance, call 1-800-HELP-MAP, or write:

USGS Information Services
Box 25286
Denver, CO 80225

For additional information, contact any USGS Earth Science Information Center or call 1-800-USA-MAPS.

This page intentionally left blank.

12 SUBMITTING OFFSITE CONSEQUENCE ANALYSIS INFORMATION FOR RISK MANAGEMENT PLAN

For the offsite consequence analysis (OCA) component of the RMP you must provide information on your worst-case and alternative release scenario(s) for toxic and flammable regulated chemicals held above the threshold quantity. The requirements for what information you must submit differ if your source has Program 1, Program 2, or Program 3 processes.

If your source has Program 1 processes, you must submit information on a worst-case release scenario for each Program 1 process. If your source has Program 2 or Program 3 processes, you must provide information on one worst-case release for all toxic regulated substances present above the threshold quantity and one worst-case release scenario for all flammable regulated substances present above the threshold quantity. You may need to submit an additional worst-case scenario if a worst-case release from another part of the source would potentially affect public receptors different from those potentially affected by the initial worst-case scenario(s) for flammable and toxic regulated substances.

In addition to a worst-case release scenario, sources with Program 2 and Program 3 processes must also provide information on alternative release scenarios. Alternative releases are releases that could occur, other than the worst-case, that may result in concentrations, overpressures, or radiant heat that reach endpoints offsite. You must present information on one alternative release scenario for each regulated toxic substance, including the substance used for the worst-case release, held above the threshold quantity and one alternative release scenario to represent all flammable substances held above the threshold quantity. The types of documentation to submit are presented below for worst-case scenarios involving toxic substances, alternative scenarios involving toxic substances, worst-case scenarios involving flammable substances, and alternative scenarios involving flammable substances.

12.1 RMP Data Required for Worst-Case Scenarios for Toxic Substances

For worst-case scenarios involving toxic substances, you will have to submit the following information. See the RMP*eSubmit User Manual for complete instructions.

- Chemical name;
- Percentage weight of the regulated liquid toxic substance (if present in a mixture);

- Physical state of the chemical released (gas, liquid, refrigerated gas, gas liquefied by pressure);
- Model used (OCA or industry-specific guidance reference tables or modeling; name of other model used);
- Scenario (gas release or liquid spill and vaporization);
- Quantity released (pounds);
- Release rate (pounds per minute);
- Duration of release (minutes) (10 minutes for gases; if you used OCA guidance for liquids, indicate either 10 or 60 minutes);
- Wind speed (meters per second) and stability class (1.5 meters per second and F stability unless you can show higher minimum wind speed or less stable atmosphere at all times during the last three years);
- Topography (rural or urban);
- Distance to endpoint (miles, rounded to two significant digits);
- Population within distance to endpoint (residential population rounded to two significant digits);
- Public receptors within the distance to endpoint (schools, residences, hospitals, prisons, recreation areas, commercial, office or industrial areas);
- Environmental receptors within the distance to endpoint (national or state parks, forests, or monuments; officially designated wildlife sanctuaries, preserves, or refuges; Federal wilderness areas); and
- Passive mitigation measures considered (dikes, enclosures, berms, drains, sumps, other).

12.2 RMP Data Required for Alternative Scenarios for Toxic Substances

For alternative scenarios involving toxic substances held above the threshold quantity in a Program 2 or Program 3 process, you will have to submit the following information. See the Risk Management Plan Data Elements Guide for complete instructions.

- Chemical name;
- Percentage weight of the regulated liquid toxic substance (if present in a mixture);
- Physical state of the chemical released (gas, liquid, refrigerated gas, gas liquefied by pressure);
- Model used (OCA or industry-specific guidance reference tables or modeling; name of other model used);
- Scenario (transfer hose failure, pipe leak, vessel leak, overfilling, rupture disk/relief valve, excess flow valve, other);
- Quantity released (pounds);
- Release rate (pounds per minute);
- Duration of release (minutes) (if you used OCA guidance, indicate either 10 or 60 minutes);
- Wind speed (meters per second) and stability class (3.0 meters per second and D stability if you use OCA guidance, otherwise use typical meteorological conditions at your site);
- Topography (rural or urban);
- Distance to endpoint (miles, rounded to two significant digits);
- Population within distance to endpoint (residential population rounded to two significant digits);

- Public receptors within the distance to endpoint (schools, residences, hospitals, prisons, recreation areas, commercial, office, or industrial areas);
- Environmental receptors within the distance to endpoint (national or state parks, forests, or monuments; officially designated wildlife sanctuaries, preserves, or refuges; Federal wilderness areas);
- Passive mitigation measures considered (dikes, enclosures, berms, drains, sumps, other); and
- Active mitigation measures considered (sprinkler system, deluge system, water curtain, neutralization, excess flow valve, flares, scrubbers, emergency shutdown system, other).

12.3 RMP Data Required for Worst-Case Scenarios for Flammable Substances

For worst-case scenarios involving flammable substances, you will have to submit the following information. See the Risk Management Plan Data Elements Guide for complete instructions.

- Chemical name;
- Model used (OCA or industry-specific guidance reference tables or modeling; name of other model used);
- Scenario (vapor cloud explosion);
- Quantity released (pounds);
- Endpoint used (for vapor cloud explosions use 1 psi);
- Distance to endpoint (miles, rounded to two significant digits);
- Population within distance to endpoint (residential population rounded to two significant digits);
- Public receptors within the distance to endpoint (schools, residences, hospitals, prisons, recreation areas, commercial, office, or industrial areas);
- Environmental receptors within the distance to endpoint (national or state parks, forests, or monuments, officially designated wildlife sanctuaries, preserves, or refuges, Federal wilderness areas); and
- Passive mitigation measures considered (blast walls, other).

12.4 RMP Data Required for Alternative Scenarios for Flammable Substances

For alternative scenarios involving flammable substances held above the threshold quantity in a Program 2 or Program 3 process, you will have to submit the following information. See the Risk Management Plan Data Elements Guide for complete instructions.

- Chemical name;
- Model used (OCA or industry-specific guidance reference tables or modeling; name of other model used);
- Scenario (vapor cloud explosion, fireball, BLEVE, pool fire, jet fire, vapor cloud fire, other);
- Quantity released (pounds);
- Endpoint used (for vapor cloud explosions, the endpoint is 1 psi overpressure; for a fireball the endpoint is 5 kw/m^2 for 40 seconds. A lower flammability limit (expressed as a percentage) may be listed as specified in NFPA documents or other generally recognized sources; these are listed in the OCA Guidance);
- Distance to endpoint (miles, rounded to two significant digits);

- Population within distance to endpoint (residential population rounded to two significant digits);
- Public receptors within the distance to endpoint (schools, residences, hospitals, prisons, recreation areas, commercial, office, or industrial areas);
- Environmental receptors within the distance to endpoint (national or state parks, forests, or monuments, officially designated wildlife sanctuaries, preserves, or refuges, Federal wilderness areas);
- Passive mitigation measures considered (e.g., dikes, fire walls, blast walls, enclosures, other); and
- Active mitigation measures considered (e.g., sprinkler system, deluge system, water curtain, excess flow valve, other).

12.5 Submitting RMPs

EPA has made RMP*eSubmit available to complete and file your RMP. RMP*eSubmit does the following:

- Provides a user-friendly, web-based RMP Submission System via EPA's secure Central Data Exchange (CDX).
- Performs data quality checks, accepts limited graphics, and provides on-line help including defining data elements and providing instructions.
- Online reporting simplifies the process. It saves you time, and improves data quality and security.
- EPA uses industry-standard technology, including encryption used by most commercial banks, as well as stringent user ID and password protocols to protect your information.
- You will be able to access your RMP online at any time.
- For a facility to submit their RMP, the certifier will first be required to set up a CDX account (or use their current one if the facility already submits data via CDX). Instructions for obtaining a CDX account and using RMP*eSubmit can be found at: www.epa.gov/emergencies/rmp.
- Upon approval of the document in submission, a confirmation e-mail will be sent to the certifier.

12.6 Other Required Documentation

Besides the information you are required to submit in your RMP, you must maintain other records of your offsite consequence analysis on site. Under 40 CFR 68.39, you must maintain the following records:

- For worst-case scenarios, a description of the vessel or pipeline and substance selected as the worst case, the assumptions and parameters used, and the rationale for selection. Assumptions include any administrative controls and any passive mitigation systems that were used to limit the quantity that could be released. You must document that anticipated effects of these controls and systems on the release quantity and rate.

- For alternative release scenarios, a description of the scenarios identified the assumptions and parameters used, and the rationale for selection of the specific scenarios. Assumptions include any administrative controls and any passive mitigation systems that were used to limit the quantity that could be released. You must document that anticipated effects of these controls and systems on the release quantity and rate.
- Documentation of estimated quantity released, release rate, and duration of the release.
- Methodology used to determine distance to an endpoint.
- Data used to estimate populations and environmental receptors potentially affected.

You are required to maintain these records for five years.

This page intentionally left blank.

APPENDIX A

REFERENCES FOR CONSEQUENCE ANALYSIS METHODS

APPENDIX A REFERENCES FOR CONSEQUENCE ANALYSIS METHODS

Exhibit A-1 lists references that may provide useful information for modeling or calculation methods that could be used in the offsite consequence analyses. This exhibit is not intended to be a complete listing of references that may be used in the consequence analysis; any appropriate model or method may be used.

Exhibit A-1
Selected References for Information on Consequence Analysis Methods

Center for Process Safety of the American Institute of Chemical Engineers (AIChE). *Guidelines for Evaluating the Characteristics of Vapor Cloud Explosions, Flash Fires, and BLEVEs*. New York: AIChE, 1994.

Center for Process Safety of the American Institute of Chemical Engineers (AIChE). *Guidelines for Use of Vapor Cloud Dispersion Models*, Second Ed. New York: AIChE, 1996.

Center for Process Safety of the American Institute of Chemical Engineers (AIChE). *International Conference and Workshop on Modeling and Mitigating the Consequences of Accidental Releases of Hazardous Materials*, September 26-29, 1995. New York: AIChE, 1995.

Federal Emergency Management Agency, U.S. Department of Transportation, U.S. Environmental Protection Agency. *Handbook of Chemical Hazard Analysis Procedures*. 1989.

Madsen, Warren W. and Robert C. Wagner. "An Accurate Methodology for Modeling the Characteristics of Explosion Effects." *Process Safety Progress*, 13 (July 1994), 171-175.

Mercx, W.P.M., D.M. Johnson, and J. Puttock. "Validation of Scaling Techniques for Experimental Vapor Cloud Explosion Investigations." *Process Safety Progress*, 14 (April 1995), 120.

Mercx, W.P.M., R.M.M. van Wees, and G. Opschoor. "Current Research at TNO on Vapor Cloud Explosion Modelling." *Process Safety Progress*, 12 (October 1993), 222.

Prugh, Richard W. "Quantitative Evaluation of Fireball Hazards." *Process Safety Progress*, 13 (April 1994), 83-91.

Scheuermann, Klaus P. "Studies About the Influence of Turbulence on the Course of Explosions." *Process Safety Progress*, 13 (October 1994), 219.

TNO Bureau for Industrial Safety, Netherlands Organization for Applied Scientific Research. *Methods for the Calculation of the Physical Effects*. The Hague, the Netherlands: Committee for the Prevention of Disasters, 1997.

TNO Bureau for Industrial Safety, Netherlands Organization for Applied Scientific Research. *Methods for the Calculation of the Physical Effects of the Escape of Dangerous Material (Liquids and Gases)*. Voorburg, the Netherlands: TNO (Commissioned by Directorate-General of Labour), 1980.

TNO Bureau for Industrial Safety, Netherlands Organization for Applied Scientific Research. *Methods for the Calculation of the Physical Effects Resulting from Releases of Hazardous Materials*. Rijswijk, the Netherlands: TNO (Commissioned by Directorate-General of Labour), 1992.

TNO Bureau for Industrial Safety, Netherlands Organization for Applied Scientific Research. *Methods for the Determination of Possible Damage to People and Objects Resulting from Releases of*

Hazardous Materials. Rijswijk, the Netherlands: TNO (Commissioned by Directorate-General of Labour), 1992.

Touma, Jawad S., et al. "Performance Evaluation of Dense Gas Dispersion Models." *Journal of Applied Meteorology*, 34 (March 1995), 603-615.

U.S. Environmental Protection Agency, Federal Emergency Management Agency, U.S. Department of Transportation. *Technical Guidance for Hazards Analysis, Emergency Planning for Extremely Hazardous Substances*. December 1987.

U.S. Environmental Protection Agency, Office of Air Quality Planning and Standards. *Workbook of Screening Techniques for Assessing Impacts of Toxic Air Pollutants*. EPA-450/4-88-009. September 1988.

U.S. Environmental Protection Agency, Office of Air Quality Planning and Standards. *Guidance on the Application of Refined Dispersion Models for Hazardous/Toxic Air Release*. EPA-454/R-93-002. May 1993.

U.S. Environmental Protection Agency, Office of Pollution Prevention and Toxic Substances. *Flammable Gases and Liquids and Their Hazards*. EPA 744-R-94-002. February 1994.

APPENDIX B

TOXIC SUBSTANCES

APPENDIX B TOXIC SUBSTANCES

B.1 Data for Toxic Substances

The exhibits in this section of Appendix B provide the data needed to carry out the calculations for regulated toxic substances using the methods presented in the text of this guidance. Exhibit B-1 presents data for toxic gases, Exhibit B-2 presents data for toxic liquids, and Exhibit B-3 presents data for several toxic substances commonly found in water solution and for oleum. Exhibit B-4 provides temperature correction factors that can be used to correct the release rates estimated for pool evaporation of toxic liquids that are released at temperatures between 25 °C to 50 °C.

The derivation of the factors presented in Exhibits B-1 - B-4 is discussed in Appendix D. The data used to develop the factors in Exhibits B-1 and B-2 are primarily from Design Institute for Physical Property Data (DIPPR), American Institute of Chemical Engineers, *Physical and Thermodynamic Properties of Pure Chemicals, Data Compilation*. Other sources, including the National Library of Medicine's Hazardous Substances Databank (HSDB) and the *Kirk-Othmer Encyclopedia of Chemical Technology*, were used for Exhibits B-1 and B-2 if data were not available from the DIPPR compilation. The factors in Exhibit B-3 were developed using data primarily from *Perry's Chemical Engineers' Handbook* and the *Kirk-Othmer Encyclopedia of Chemical Technology*. The temperature correction factors in Exhibit B-4 were developed using vapor pressure data derived from the vapor pressure coefficients in the DIPPR compilation.

Exhibit B-1
Data for Toxic Gases

CAS Number	Chemical Name	Molecular Weight	Ratio of Specific Heats	Toxic Endpoint[a]			Liquid Factor Boiling (LFB)	Density Factor (DF) (Boiling)	Gas Factor (GF)[g]	Vapor Pressure @25 °C (psia)	Reference Table[b]
				mg/L	ppm	Basis					
7664-41-7	Ammonia (anhydrous)[c]	17.03	1.31	0.14	200	ERPG-2	0.073	0.71	14	145	Buoyant[d]
7784-42-1	Arsine	77.95	1.28	0.0019	0.6	EHS-LOC (IDLH)	0.23	0.30	30	239	Dense
10294-34-5	Boron trichloride	117.17	1.15	0.010	2	EHS-LOC (Tox[e])	0.22	0.36	36	22.7	Dense
7637-07-2	Boron trifluoride	67.81	1.20	0.028	10	EHS-LOC (IDLH)	0.25	0.31	28	f	Dense
7782-50-5	Chlorine	70.91	1.32	0.0087	3	ERPG-2	0.19	0.31	29	113	Dense
10049-04-4	Chlorine dioxide	67.45	1.25	0.0028	1	EHS-LOC equivalent (IDLH)[g]	0.15	0.30	28	24.3	Dense
506-77-4	Cyanogen chloride	61.47	1.22	0.030	12	EHS-LOC equivalent (Tox)[h]	0.14	0.41	26	23.7	Dense
19287-45-7	Diborane	27.67	1.17	0.0011	1	ERPG-2	0.13	1.13	17	f	Buoyant[d]
75-21-8	Ethylene oxide	44.05	1.21	0.090	50	ERPG-2	0.12	0.55	22	25.4	Dense
7782-41-4	Fluorine	38.00	1.36	0.0039	2.5	EHS-LOC (IDLH)	0.35	0.32	22	f	Dense
50-00-0	Formaldehyde (anhydrous)[c]	30.03	1.31	0.012	10	ERPG-2	0.10	0.59	19	75.2	Dense
74-90-8	Hydrocyanic acid	27.03	1.30	0.011	10	ERPG-2	0.079	0.72	18	14.8	Buoyant[d]
7647-01-0	Hydrogen chloride (anhydrous)[c]	36.46	1.40	0.030	20	ERPG-2	0.15	0.41	21	684	Dense
7664-39-3	Hydrogen fluoride (anhydrous)[c]	20.01	1.40	0.016	20	ERPG-2	0.066	0.51	16	17.7	Buoyant[i]
7783-07-5	Hydrogen selenide	80.98	1.32	0.00066	0.2	EHS-LOC (IDLH)	0.21	0.25	31	151	Dense
7783-06-4	Hydrogen sulfide	34.08	1.32	0.042	30	ERPG-2	0.13	0.51	20	302	Dense
74-87-3	Methyl chloride	50.49	1.26	0.82	400	ERPG-2	0.14	0.48	24	83.2	Dense
74-93-1	Methyl mercaptan	48.11	1.20	0.049	25	ERPG-2	0.12	0.55	23	29.2	Dense
10102-43-9	Nitric oxide	30.01	1.38	0.031	25	EHS-LOC (TLV[k])	0.21	0.38	19	f	Dense
75-44-5	Phosgene	98.92	1.17	0.00081	0.2	ERPG-2	0.20	0.35	33	27.4	Dense

Exhibit B-1 (continued)

CAS Number	Chemical Name	Molecular Weight	Ratio of Specific Heats	Toxic Endpoint[a]			Liquid Factor Boiling (LFB)	Density Factor (DF) (Boiling)	Gas Factor (GF)[k]	Vapor Pressure @25 °C (psia)	Reference Table[b]
				mg/L	ppm	Basis					
7803-51-2	Phosphine	34.00	1.29	0.0035	2.5	ERPG-2	0.15	0.66	20	567	Dense
7446-09-5	Sulfur dioxide (anhydrous)	64.07	1.26	0.0078	3	ERPG-2	0.16	0.33	27	58.0	Dense
7783-60-0	Sulfur tetrafluoride	108.06	1.30	0.0092	2	EHS-LOC (Tox[e])	0.25	0.25 (at -73 °C)	36	293	Dense

Notes:

[a] Toxic endpoints are specified in Appendix A to 40 CFR part 68 in units of mg/L. To convert from units of mg/L to mg/m^3, multiply by 1,000. To convert mg/L to ppm, use the following equation:

$$Endpoint_{ppm} = \frac{Endpoint_{mg/L} \times 1{,}000 \times 24.5}{Molecular\ Weight}$$

[b] "Buoyant" in the Reference Table column refers to the tables for neutrally buoyant gases and vapors; "Dense" refers to the tables for dense gases and vapors. See Appendix D, Section D.4.4, for more information on the choice of reference tables.
[c] See Exhibit B-3 of this appendix for data on water solutions.
[d] Gases that are lighter than air may behave as dense gases upon release if liquefied under pressure or cold; consider the conditions of release when choosing the appropriate table.
[e] LOC is based on the IDLH-equivalent level estimated from toxicity data.
[f] Cannot be liquefied at 25 °C.
[g] Not an EHS; LOC-equivalent value was estimated from one-tenth of the IDLH.
[h] Not an EHS; LOC-equivalent value was estimated from one-tenth of the IDLH-equivalent level estimated from toxicity data.
[i] Hydrogen fluoride is lighter than air, but may behave as a dense gas upon release under some circumstances (e.g., release under pressure, high concentration in the released cloud) because of hydrogen bonding; consider the conditions of release when choosing the appropriate table.
[j] LOC based on Threshold Limit Value (TLV) - Time-weighted average (TWA) developed by the American Conference of Governmental Industrial Hygienists (ACGIH).
[k] Use GF for gas leaks under choked (maximum) flow conditions.

Exhibit B-2
Data for Toxic Liquids

CAS Number	Chemical Name	Molecular Weight	Vapor Pressure at 25 °C (mm Hg)	Toxic Endpoint[a]			Liquid Factors		Density Factor (DF)	Liquid Leak Factor (LLF)[c]	Reference Table[b]	
				mg/L	ppm	Basis	Ambient (LFA)	Boiling (LFB)			Worst Case	Alternative Case
107-02-8	Acrolein	56.06	274	0.0011	0.5	ERPG-2	0.047	0.12	0.58	40	Dense	Dense
107-13-1	Acrylonitrile	53.06	108	0.076	35	ERPG-2	0.018	0.11	0.61	39	Dense	Dense
814-68-6	Acrylyl chloride	90.51	110	0.00090	0.2	EHS-LOC (Tox[e])	0.026	0.15	0.44	54	Dense	Dense
107-18-6	Allyl alcohol	58.08	26.1	0.036	15	EHS-LOC (IDLH)	0.0046	0.11	0.58	41	Dense	Buoyant[d]
107-11-9	Allylamine	57.10	242	0.0032	1	EHS-LOC (Tox[e])	0.042	0.12	0.64	36	Dense	Dense
7784-34-1	Arsenous trichloride	181.28	10	0.01	1	EHS-LOC (Tox[e])	0.0037	0.21	0.23	100	Dense	Buoyant[d]
353-42-4	Boron trifluoride compound with methyl ether (1:1)	113.89	11	0.023	5	EHS-LOC (Tox[e])	0.0030	0.16	0.49	48	Dense	Buoyant[d]
7726-95-6	Bromine	159.81	212	0.0065	1	ERPG-2	0.073	0.23	0.16	150	Dense	Dense
75-15-0	Carbon disulfide	76.14	359	0.16	50	ERPG-2	0.075	0.15	0.39	60	Dense	Dense
67-66-3	Chloroform	119.38	196	0.49	100	EHS-LOC (IDLH)	0.055	0.19	0.33	71	Dense	Dense
542-88-1	Chloromethyl ether	114.96	29.4	0.00025	0.05	EHS-LOC (Tox[e])	0.0080	0.17	0.37	63	Dense	Dense
107-30-2	Chloromethyl methyl ether	80.51	199	0.0018	0.6	EHS-LOC (Tox[e])	0.043	0.15	0.46	51	Dense	Dense
4170-30-3	Crotonaldehyde	70.09	33.1	0.029	10	ERPG-2	0.0066	0.12	0.58	41	Dense	Buoyant[d]
123-73-9	Crotonaldehyde, (E)-	70.09	33.1	0.029	10	ERPG-2	0.0066	0.12	0.58	41	Dense	Buoyant[d]
108-91-8	Cyclohexylamine	99.18	10.1	0.16	39	EHS-LOC (Tox[e])	0.0025	0.14	0.56	41	Dense	Buoyant[d]
75-78-5	Dimethyldichlorosilane	129.06	141	0.026	5	ERPG-2	0.042	0.20	0.46	51	Dense	Dense
57-14-7	1,1-Dimethylhydrazine	60.10	157	0.012	5	EHS-LOC (IDLH)	0.028	0.12	0.62	38	Dense	Dense
106-89-8	Epichlorohydrin	92.53	17.0	0.076	20	ERPG-2	0.0040	0.14	0.42	57	Dense	Buoyant[d]
107-15-3	Ethylenediamine	60.10	12.2	0.49	200	EHS-LOC (IDLH)	0.0022	0.13	0.54	43	Dense	Buoyant[d]
151-56-4	Ethyleneimine	43.07	211	0.018	10	EHS-LOC (IDLH)	0.030	0.10	0.58	40	Dense	Dense
110-00-9	Furan	68.08	600	0.0012	0.4	EHS-LOC (Tox[e])	0.12	0.14	0.52	45	Dense	Dense
302-01-2	Hydrazine	32.05	14.4	0.011	8	EHS-LOC (IDLH)	0.0017	0.069	0.48	48	Buoyant[d]	Buoyant[d]

Exhibit B-2 (continued)

CAS Number	Chemical Name	Molecular Weight	Vapor Pressure at 25 °C (mm Hg)	Toxic Endpoint[a]			Liquid Factors		Density Factor (DF)	Liquid Leak Factor (LLF)[d]	Reference Table[b]	
				mg/L	ppm	Basis	Ambient (LFA)	Boiling (LFB)			Worst Case	Alternative Case
13463-40-6	Iron, pentacarbonyl-	195.90	40	0.00044	0.05	EHS-LOC (Tox[c])	0.016	0.24	0.33	70	Dense	Dense
78-82-0	Isobutyronitrile	69.11	32.7	0.14	50	ERPG-2	0.0064	0.12	0.63	37	Dense	Buoyant[d]
108-23-6	Isopropyl chloroformate	122.55	28	0.10	20	EHS-LOC (Tox[c])	0.0080	0.17	0.45	52	Dense	Dense
126-98-7	Methacrylonitrile	67.09	71.2	0.0027	1	EHS-LOC (TLV[e])	0.014	0.12	0.61	38	Dense	Dense
79-22-1	Methyl chloroformate	94.50	108	0.0019	0.5	EHS-LOC (Tox[c])	0.026	0.16	0.40	58	Dense	Dense
60-34-4	Methyl hydrazine	46.07	49.6	0.0094	5	EHS-LOC (IDLH)	0.0074	0.094	0.56	42	Dense	Buoyant[d]
624-83-9	Methyl isocyanate	57.05	457	0.0012	0.5	ERPG-2	0.079	0.13	0.52	45	Dense	Dense
556-64-9	Methyl thiocyanate	73.12	10	0.085	29	EHS-LOC (Tox[c])	0.0020	0.11	0.45	51	Dense	Buoyant[d]
75-79-6	Methyltrichlorosilane	149.48	173	0.018	3	ERPG-2	0.057	0.22	0.38	61	Dense	Dense
13463-39-3	Nickel carbonyl	170.73	400	0.00067	0.1	EHS-LOC (Tox[c])	0.14	0.26	0.37	63	Dense	Dense
7697-37-2	Nitric acid (100%)[f]	63.01	63.0	0.026	10	EHS-LOC (Tox[c])	0.012	0.12	0.32	73	Dense	Dense
79-21-0	Peracetic acid	76.05	13.9	0.0045	1.5	EHS-LOC (Tox[c])	0.0029	0.12	0.40	58	Dense	Buoyant[d]
594-42-3	Perchloromethylmercaptan	185.87	6	0.0076	1	EHS-LOC (IDLH)	0.0023	0.20	0.29	81	Dense	Buoyant[d]
10025-87-3	Phosphorus oxychloride	153.33	35.8	0.0030	0.5	EHS-LOC (Tox[c])	0.012	0.20	0.29	80	Dense	Dense
7719-12-2	Phosphorus trichloride	137.33	120	0.028	5	EHS-LOC (IDLH)	0.037	0.20	0.31	75	Dense	Dense
110-89-4	Piperidine	85.15	32.1	0.022	6	EHS-LOC (Tox[c])	0.0072	0.13	0.57	41	Dense	Buoyant[d]
107-12-0	Propionitrile	55.08	47.3	0.0037	1.6	EHS-LOC (Tox[c])	0.0080	0.10	0.63	37	Dense	Buoyant[d]
109-61-5	Propyl chloroformate	122.56	20.0	0.010	2	EHS-LOC (Tox[c])	0.0058	0.17	0.45	52	Dense	Buoyant[d]
75-55-8	Propyleneimine	57.10	187	0.12	50	EHS-LOC (IDLH)	0.032	0.12	0.61	39	Dense	Dense
75-56-9	Propylene oxide	58.08	533	0.59	250	ERPG-2	0.093	0.13	0.59	40	Dense	Dense
7446-11-9	Sulfur trioxide	80.06	263	0.010	3	ERPG-2	0.057	0.15	0.26	91	Dense	Dense
75-74-1	Tetramethyllead	267.33	22.5	0.0040	0.4	EHS-LOC (IDLH)	0.011	0.29	0.24	96	Dense	Dense
509-14-8	Tetranitromethane	196.04	11.4	0.0040	0.5	EHS-LOC (IDLH)	0.0045	0.22	0.30	78	Dense	Buoyant[d]

Exhibit B-2 (continued)

CAS Number	Chemical Name	Molecular Weight	Vapor Pressure at 25 °C (mm Hg)	Toxic Endpoint[a]			Liquid Factors		Density Factor (DF)	Liquid Leak Factor (LLF)[i]	Reference Table[b]	
				mg/L	ppm	Basis	Ambient (LFA)	Boiling (LFB)			Worst Case	Alternative Case
7550-45-0	Titanium tetrachloride	189.69	12.4	0.020	2.6	ERPG-2	0.0048	0.21	0.28	82	Dense	Buoyant[d]
584-84-9	Toluene 2,4-diisocyanate	174.16	0.017	0.0070	1	EHS-LOC (IDLH)	0.000006	0.16	0.40	59	Buoyant[d]	Buoyant[d]
91-08-7	Toluene 2,6-diisocyanate	174.16	0.05	0.0070	1	EHS-LOC (IDLH[g])	0.000018	0.16	0.40	59	Buoyant[d]	Buoyant[d]
26471-62-5	Toluene diisocyanate (unspecified isomer)	174.16	0.017	0.0070	1	EHS-LOC equivalent (IDLH[h])	0.000006	0.16	0.40	59	Buoyant[d]	Buoyant[d]
75-77-4	Trimethylchlorosilane	108.64	231	0.050	11	EHS-LOC (Tox[c])	0.061	0.18	0.57	41	Dense	Dense
108-05-4	Vinyl acetate monomer	86.09	113	0.26	75	ERPG-2	0.026	0.15	0.53	45	Dense	Dense

Notes:

[a] Toxic endpoints are specified in the Appendix A to 40 CFR part 68 in units of mg/L. To convert from units of mg/L to mg/m^3, multiply by 1,000. To convert mg/L to ppm, use the following equation:

$$Endpoint_{ppm} = \frac{Endpoint_{mg/L} \times 1{,}000 \times 24.5}{Molecular\ Weight}$$

[b] "Buoyant" in the Reference Table column refers to the tables for neutrally buoyant gases and vapors; "Dense" refers to the tables for dense gases and vapors. See Appendix D, Section D.4.4, for more information on the choice of reference tables.

[c] LOC is based on IDLH-equivalent level estimated from toxicity data.

[d] Use dense gas table if substance is at an elevated temperature.

[e] LOC based on Threshold Limit Value (TLV) - Time-weighted average (TWA) developed by the American Conference of Governmental Industrial Hygienists (ACGIH).

[f] See Exhibit B-3 of this appendix for data on water solutions.

[g] LOC for this isomer is based on IDLH for toluene 2,4-diisocyanate.

[h] Not an EHS; LOC-equivalent value is based on IDLH for toluene 2,4-diisocyanate.

[i] Use the LLF only for leaks from tanks at atmospheric pressure.

Exhibit B-3
Data for Water Solutions of Toxic Substances and for Oleum For Wind Speeds of 1.5 and 3.0 Meters per Second (m/s)

CAS Number	Regulated Substance in Solution	Molecular Weight	Toxic Endpoint[a]			Initial Concentration (Wt %)	10-min. Average Vapor Pressure (mm Hg)		Liquid Factor at 25° C (LFA)		Density Factor (DF)	Liquid Leak Factor (LLF)	Reference Table[b]	
			mg/L	ppm	Basis		1.5 m/s	3.0 m/s	1.5 m/s	3.0 m/s			Worst	Alternative
7664-41-7	Ammonia	17.03	0.14	200	ERPG-2	30	332	248	0.026	0.019	0.55	43	Buoyant	Buoyant
						24	241	184	0.019	0.014	0.54	44	Buoyant	Buoyant
						20	190	148	0.015	0.011	0.53	44	Buoyant	Buoyant
50-00-0	Formaldehyde	30.027	0.012	10	ERPG-2	37	1.5	1.4	0.0002	0.0002	0.44	53	Buoyant	Buoyant
7647-01-0	Hydrochloric acid	36.46	0.030	20	ERPG-2	38	78	55	0.010	0.0070	0.41	57	Dense	Buoyant[d]
						37	67	48	0.0085	0.0062	0.42	57	Dense	Buoyant[d]
						36[c]	56	42	0.0072	0.0053	0.42	57	Dense	Buoyant[d]
						34[c]	38	29	0.0048	0.0037	0.42	56	Dense	Buoyant[d]
						30[c]	13	12	0.0016	0.0015	0.42	55	Buoyant[d]	Buoyant[d]
7664-39-3	Hydrofluoric acid	20.01	0.016	20	ERPG-2	70	124	107	0.011	0.010	0.39	61	Buoyant	Buoyant
						50	16	15	0.0014	0.0013	0.41	58	Buoyant	Buoyant
7697-37-2	Nitric acid	63.01	0.026	10	EHS-LOC (IDLH)	90	25	22	0.0046	0.0040	0.33	71	Dense	Buoyant[d]
						85	17	16	0.0032	0.0029	0.33	70	Dense	Buoyant[d]
						80	10.2	10	0.0019	0.0018	0.33	70	Dense	Buoyant[d]
8014-95-7	Oleum - based on SO_3	80.06 (SO_3)	0.010	3	ERPG-2	30 (SO_3)	3.5 (SO_3)	3.4 (SO_3)	0.0008	0.0007	0.25	93	Buoyant[d]	Buoyant[d]

Notes:

[a] Toxic endpoints are specified in the Appendix A to 40 CFR part 68 in units of mg/L. See Notes to Exhibit B-1 or B-2 for converting to other units.

[b] "Buoyant" in the Reference Table column refers to the tables for neutrally buoyant gases and vapors; "Dense" refers to the tables for dense gases and vapors. See Appendix D, Section D.4.4, for more information on the choice of reference tables.

[c] Hydrochloric acid in concentrations below 37 percent is not regulated.

[d] Use dense gas table if substance is at an elevated temperature.

Exhibit B-4
Temperature Correction Factors for Liquids Evaporating from Pools at Temperatures Between 25 °C and 50 °C (77 °F and 122 °F)

CAS Number	Chemical Name	Boiling Point (°C)	Temperature Correction Factor (TCF)				
			30 °C (86 °F)	35 °C (95 °F)	40 °C (104 °F)	45 °C (113 °F)	50 °C (122 °F)
107-02-8	Acrolein	52.69	1.2	1.4	1.7	2.0	2.3
107-13-1	Acrylonitrile	77.35	1.2	1.5	1.8	2.1	2.5
814-68-6	Acrylyl chloride	75.00	ND	ND	ND	ND	ND
107-18-6	Allyl alcohol	97.08	1.3	1.7	2.2	2.9	3.6
107-11-9	Allylamine	53.30	1.2	1.5	1.8	2.1	2.5
7784-34-1	Arsenous trichloride	130.06	ND	ND	ND	ND	ND
353-42-4	Boron trifluoride compound with methyl ether (1:1)	126.85	ND	ND	ND	ND	ND
7726-95-6	Bromine	58.75	1.2	1.5	1.7	2.1	2.5
75-15-0	Carbon disulfide	46.22	1.2	1.4	1.6	1.9	LFB
67-66-3	Chloroform	61.18	1.2	1.5	1.8	2.1	2.5
542-88-1	Chloromethyl ether	104.85	1.3	1.6	2.0	2.5	3.1
107-30-2	Chloromethyl methyl ether	59.50	1.2	1.5	1.8	2.1	2.5
4170-30-3	Crotonaldehyde	104.10	1.3	1.6	2.0	2.5	3.1
123-73-9	Crotonaldehyde, (E)-	102.22	1.3	1.6	2.0	2.5	3.1
108-91-8	Cyclohexylamine	134.50	1.3	1.7	2.1	2.7	3.4
75-78-5	Dimethyldichlorosilane	70.20	1.2	1.5	1.8	2.1	2.5
57-14-7	1,1-Dimethylhydrazine	63.90	ND	ND	ND	ND	ND
106-89-8	Epichlorohydrin	118.50	1.3	1.7	2.1	2.7	3.4
107-15-3	Ethylenediamine	36.26	1.3	1.8	LFB	LFB	LFB
151-56-4	Ethyleneimine	55.85	1.2	1.5	1.8	2.2	2.7
110-00-9	Furan	31.35	1.2	LFB	LFB	LFB	LFB
302-01-2	Hydrazine	113.50	1.3	1.7	2.2	2.9	3.6
13463-40-6	Iron, pentacarbonyl-	102.65	ND	ND	ND	ND	ND
78-82-0	Isobutyronitrile	103.61	1.3	1.6	2.0	2.5	3.1
108-23-6	Isopropyl chloroformate	104.60	ND	ND	ND	ND	ND
126-98-7	Methacrylonitrile	90.30	1.2	1.5	1.8	2.2	2.6
79-22-1	Methyl chloroformate	70.85	1.3	1.6	1.9	2.4	2.9
60-34-4	Methyl hydrazine	87.50	ND	ND	ND	ND	ND
624-83-9	Methyl isocyanate	38.85	1.2	1.4	LFB	LFB	LFB
556-64-9	Methyl thiocyanate	130.00	ND	ND	ND	ND	ND
75-79-6	Methyltrichlorosilane	66.40	1.2	1.4	1.7	2.0	2.4

Exhibit B-4 (continued)

CAS Number	Chemical Name	Boiling Point (°C)	Temperature Correction Factor (TCF)				
			30 °C (86 °F)	35 °C (95 °F)	40 °C (104 °F)	45 °C (113 °F)	50 °C (122 °F)
13463-39-3	Nickel carbonyl	42.85	ND	ND	ND	ND	ND
7697-37-2	Nitric acid	83.00	1.3	1.6	2.0	2.5	3.1
79-21-0	Peracetic acid	109.85	1.3	1.8	2.3	3.0	3.8
594-42-3	Perchloromethylmercaptan	147.00	ND	ND	ND	ND	ND
10025-87-3	Phosphorus oxychloride	105.50	1.3	1.6	1.9	2.4	2.9
7719-12-2	Phosphorus trichloride	76.10	1.2	1.5	1.8	2.1	2.5
110-89-4	Piperidine	106.40	1.3	1.6	2.0	2.4	3.0
107-12-0	Propionitrile	97.35	1.3	1.6	1.9	2.3	2.8
109-61-5	Propyl chloroformate	112.40	ND	ND	ND	ND	ND
75-55-8	Propyleneimine	60.85	1.2	1.5	1.8	2.1	2.5
75-56-9	Propylene oxide	33.90	1.2	LFB	LFB	LFB	LFB
7446-11-9	Sulfur trioxide	44.75	1.3	1.7	LFB	LFB	LFB
75-74-1	Tetramethyllead	110.00	ND	ND	ND	ND	ND
509-14-8	Tetranitromethane	125.70	1.3	1.7	2.2	2.8	3.5
7550-45-0	Titanium tetrachloride	135.85	1.3	1.6	2.0	2.6	3.2
584-84-9	Toluene 2,4-diisocyanate	251.00	1.6	2.4	3.6	5.3	7.7
91-08-7	Toluene 2,6-diisocyanate	244.85	ND	ND	ND	ND	ND
26471-62-5	Toluene diisocyanate (unspecified isomer)	250.00	1.6	2.4	3.6	5.3	7.7
75-77-4	Trimethylchlorosilane	57.60	1.2	1.4	1.7	2.0	2.3
108-05-4	Vinyl acetate monomer	72.50	1.2	1.5	1.9	2.3	2.7

Notes:

ND: No data available.
LFB: Chemical above boiling point at this temperature; use LFB for analysis.

B.2 Mixtures Containing Toxic Liquids

In case of a spill of a liquid mixture containing a regulated toxic substance (with the exception of common water solutions, discussed in Section 3.3 in the text), the area of the pool formed by the entire liquid spill is determined as described in Section 3.2.2 or 3.2.3. For the area determination, if the density of the mixture is unknown, the density of the regulated substance in the mixture may be assumed as the density of the entire mixture.

If the partial vapor pressure of the regulated substance in the mixture is known, that vapor pressure may be used to derive a release rate using the equations in Section 3.2. If the partial vapor pressure of the regulated toxic substance in the mixture is unknown, it may be estimated from the vapor pressure of the pure substance (listed in Exhibit B-2, Appendix B) and the concentration in the mixture, if you assume the mixture is an ideal solution, where an ideal solution is one in which there is complete uniformity of cohesive forces. This method may overestimate or underestimate the partial pressure for a regulated substance that interacts with the other components of a mixture or solution. For example, water solutions are generally not ideal. This method is likely to overestimate the partial pressure of regulated substances in water solution if there is hydrogen bonding in the solution (e.g., solutions of acids or alcohols in water).

To estimate partial pressure for a regulated substance in a mixture or solution, use the following steps, based on Raoult's Law for ideal solutions:

- Determine the mole fraction of the regulated substance in the mixture.
  - -- The mole fraction of the regulated substance in the mixture is the number of moles of the regulated substance in the mixture divided by the total number of moles of all substances in the mixture.
  - -- If the molar concentration (moles per liter) of each component of the mixture is known, the mole fraction may be determined as follows:

$$X_r = \frac{M_r \times V_t}{\sum_{i=1}^{n} (M_i \times V_t)} \quad \textbf{(B-1)}$$

or (canceling out V_t):

$$X_r = \frac{M_r}{\sum_{i=1}^{n} M_i} \quad \textbf{(B-2)}$$

where: X_r = Mole fraction of regulated substance in mixture (unitless)
M_r = Molar concentration of regulated substance in mixture (moles per liter)
V_t = Total volume of mixture (liters)
n = Number of components of mixture
M_i = Molar concentration of each component of mixture (moles per liter)

For a mixture with three components, this would correspond to:

$$X_r = \frac{M_r}{M_r + M_2 + M_3} \tag{B-3}$$

where: X_r = Mole fraction of regulated substance in mixture (unitless)
M_r = Molar concentration of regulated substance (first component) in mixture (moles per liter)
M_2 = Molar concentration of second component of mixture (moles per liter)
M_3 = Molar concentration of any other components of mixture (moles per liter)

-- If the weight of each of the components of the mixture is known, the mole fraction of the regulated substance in the mixture may be calculated as follows:

$$X_r = \frac{\left(\frac{W_r}{MW_r}\right)}{\sum_{i=1}^{n}\left(\frac{W_i}{MW_i}\right)} \tag{B-4}$$

where: X_r = Mole fraction of the regulated substance
W_r = Weight of the regulated substance
MW_r = Molecular weight of the regulated substance
n = Number of components of the mixture
W_i = Weight of each component of the mixture
MW_i = Molecular weight of each component of the mixture

(Note: Weights can be in any consistent units.)

For a mixture with three components, this corresponds to:

$$X_r = \frac{\left(\frac{W_r}{MW_r}\right)}{\left(\frac{W_r}{MW_r}\right) + \left(\frac{W_2}{MW_2}\right) + \left(\frac{W_3}{MW_3}\right)} \tag{B-5}$$

where: X_r = Mole fraction of the regulated substance
W_r = Weight of the regulated substance (first component of the mixture)
MW_r = Molecular weight of the regulated substance
W_2 = Weight of the second component of the mixture
MW_2 = Molecular weight of the second component of the mixture
W_3 = Weight of the third component of the mixture
MW_3 = Molecular weight of the third component of the mixture

(Note: Weights can be in any consistent units.)

- Estimate the partial vapor pressure of the regulated substance in the mixture as follows:

$$VP_m = X_r \times VP_p \tag{B-6}$$

where: VP_m = Partial vapor pressure of the regulated substance in the mixture (millimeters of mercury (mm Hg))
X_r = Mole fraction of the regulated substance (unitless)
VP_p = Vapor pressure of the regulated substance in pure form at the same temperature as the mixture (mm Hg) (vapor pressure at 25 °C is given in Exhibit B-1, Appendix B)

The evaporation rate for the regulated substance in the mixture is determined as for pure substances, with VP_m as the vapor pressure. If the mixture contains more than one regulated toxic substance, carry out the analysis individually for each of the regulated components. The release rate equation is:

$$QR = \frac{0.0035 \times U^{0.78} \times MW^{2/3} \times A \times VP}{T} \tag{B-7}$$

where: QR = Evaporation rate (pounds per minute)
U = Wind speed (meters per second)
MW = Molecular weight (given in Exhibit B-2, Appendix B)
A = Surface area of pool formed by the entire quantity of the mixture (square feet) (determined as described in 3.2.2)
VP = Vapor pressure (mm Hg) (VP_m from Equation B-4 above)
T = Temperature (Kelvin (K); temperature in °C plus 273, or 298 for 25 °C)

See Appendix D, Section D.2.1 for more discussion of the evaporation rate equation. Equation B-7 is derived from Equation D-1.

Worst-case consequence distances to the toxic endpoint may be estimated from the release rate using the tables and instructions presented in Chapter 4.

APPENDIX C

FLAMMABLE SUBSTANCES

APPENDIX C FLAMMABLE SUBSTANCES

C.1 Equation for Estimation of Distance to 1 psi Overpressure for Vapor Cloud Explosions

For a worst-case release of flammable gases and volatile flammable liquids, the release rate is not considered. The total quantity of the flammable substance is assumed to form a vapor cloud. The entire contents of the cloud is assumed to be within the flammability limits, and the cloud is assumed to explode. For the worst-case, analysis, 10 percent of the flammable vapor in the cloud is assumed to participate in the explosion (i.e., the yield factor is 0.10). Consequence distances to an overpressure level of 1 pound per square inch (psi) may be determined using the following equation, which is based on the TNT-equivalency method:

$$D = 17 \times \left(0.1 \times W_f \times \frac{HC_f}{HC_{TNT}}\right)^{1/3} \qquad \text{(C-1)}$$

where:			
	D	=	Distance to overpressure of 1 psi (meters)
	W_f	=	Weight of flammable substance (kilograms or pounds/2.2)
	Hc_f	=	Heat of combustion of flammable substance (kilojoules per kilogram) (listed in Exhibit C-1)
	HC_{TNT}	=	Heat of explosion of trinitrotoluene (TNT) (4,680 kilojoules per kilogram)

The factor 17 is a constant for damages associated with 1.0 psi overpressures. The factor 0.1 represents an explosion efficiency of 10 percent. To convert distances from meters to miles, multiply by 0.00062.

Alternatively, use the following equation for quantity in pounds and distance in miles:

$$D_{mi} = 0.0081 \times \left(0.1 \times W_{lb} \times \frac{HC_f}{HC_{TNT}}\right)^{1/3} \qquad \text{(C-2)}$$

where:			
	D_{mi}	=	Distance to overpressure of 1 psi (miles)
	W_{lb}	=	Weight of flammable substance (pounds)

These equations were used to derive Reference Table 13 for worst-case distances to the overpressure endpoint (1 psi) for vapor cloud explosions.

C.2 Mixtures of Flammable Substances

For a mixture of flammable substances, you may estimate the heat of combustion of the mixture from the heats of combustion of the components of the mixture using the equation below and then use the equation given in the previous section of this appendix to determine the vapor cloud explosion distance. The heat of combustion of the mixture may be estimated as follows:

$$HC_m = \frac{W_x}{W_m} \times HC_x + \frac{W_y}{W_m} \times HC_y \qquad \textbf{(C-3)}$$

where:

HC_m	=	Heat of combustion of mixture (kilojoules per kilogram)
W_x	=	Weight of component "X" in mixture (kilograms or pounds/2.2)
W_m	=	Total weight of mixture (kilograms or pounds/2.2)
HC_x	=	Heat of combustion of component "X" (kilojoules per kilogram)
W_y	=	Weight of component "Y" in mixture (kilograms or pounds/2.2)
HC_y	=	Heat of combustion of component "Y" (kilojoules per kilogram)

Heats of combustion for regulated flammable substances are listed in Exhibit C-1 in the next section (Section C.3) of this appendix.

C.3 Data for Flammable Substances

The exhibits in this section of Appendix C provide the data needed to carry out the calculations for regulated flammable substances using the methods presented in the text of this guidance. Exhibit C-1 presents heat of combustion data for all regulated flammable substances, Exhibit C-2 presents additional data for flammable gases, and Exhibit C-3 presents additional data for flammable liquids. The heats of combustion in Exhibit C-1 and the data used to develop the factors in Exhibits C-2 and C-3 are primarily from Design Institute for Physical Property Data, American Institute of Chemical Engineers, *Physical and Thermodynamic Properties of Pure Chemicals, Data Compilation*. The derivation of the factors presented in Exhibits C-2 and C-3 is discussed in Appendix D.

Exhibit C-1
Heats of Combustion for Flammable Substances

CAS No.	Chemical Name	Physical State at 25° C	Heat of Combustion (kjoule/kg)
75-07-0	Acetaldehyde	Gas	25,072
74-86-2	Acetylene [Ethyne]	Gas	48,222
598-73-2	Bromotrifluoroethylene [Ethene, bromotrifluoro-]	Gas	1,967
106-99-0	1,3-Butadiene	Gas	44,548
106-97-8	Butane	Gas	45,719
25167-67-3	Butene	Gas	45,200*
590-18-1	2-Butene-cis	Gas	45,171
624-64-6	2-Butene-trans [2-Butene, (E)]	Gas	45,069
106-98-9	1-Butene	Gas	45,292
107-01-7	2-Butene	Gas	45,100*
463-58-1	Carbon oxysulfide [Carbon oxide sulfide (COS)]	Gas	9,126
7791-21-1	Chlorine monoxide [Chlorine oxide]	Gas	1,011*
590-21-6	1-Chloropropylene [1-Propene, 1-chloro-]	Liquid	23,000*
557-98-2	2-Chloropropylene [1-Propene, 2-chloro-]	Gas	22,999
460-19-5	Cyanogen [Ethanedinitrile]	Gas	21,064
75-19-4	Cyclopropane	Gas	46,560
4109-96-0	Dichlorosilane [Silane, dichloro-]	Gas	8,225
75-37-6	Difluoroethane [Ethane, 1,1-difluoro-]	Gas	11,484
124-40-3	Dimethylamine [Methanamine, N-methyl-]	Gas	35,813
463-82-1	2,2-Dimethylpropane [Propane, 2,2-dimethyl-]	Gas	45,051
74-84-0	Ethane	Gas	47,509
107-00-6	Ethyl acetylene [1-Butyne]	Gas	45,565
75-04-7	Ethylamine [Ethanamine]	Gas	35,210
75-00-3	Ethyl chloride [Ethane, chloro-]	Gas	19,917
74-85-1	Ethylene [Ethene]	Gas	47,145

Exhibit C-1 (continued)

CAS No.	Chemical Name	Physical State at 25° C	Heat of Combustion (kjoule/kg)
60-29-7	Ethyl ether [Ethane, 1,1'-oxybis-]	Liquid	33,775
75-08-1	Ethyl mercaptan [Ethanethiol]	Liquid	27,948
109-95-5	Ethyl nitrite [Nitrous acid, ethyl ester]	Gas	18,000
1333-74-0	Hydrogen	Gas	119,950
75-28-5	Isobutane [Propane, 2-methyl]	Gas	45,576
78-78-4	Isopentane [Butane, 2-methyl-]	Liquid	44,911
78-79-5	Isoprene [1,3-Butadiene, 2-methyl-]	Liquid	43,809
75-31-0	Isopropylamine [2-Propanamine]	Liquid	36,484
75-29-6	Isopropyl chloride [Propane, 2-chloro-]	Liquid	23,720
74-82-8	Methane	Gas	50,029
74-89-5	Methylamine [Methanamine]	Gas	31,396
563-45-1	3-Methyl-1-butene	Gas	44,559
563-46-2	2-Methyl-1-butene	Liquid	44,414
115-10-6	Methyl ether [Methane, oxybis-]	Gas	28,835
107-31-3	Methyl formate [Formic acid, methyl ester]	Liquid	15,335
115-11-7	2-Methylpropene [1-Propene, 2-methyl-]	Gas	44,985
504-60-9	1,3-Pentadiene	Liquid	43,834
109-66-0	Pentane	Liquid	44,697
109-67-1	1-Pentene	Liquid	44,625
646-04-8	2-Pentene, (E)-	Liquid	44,458
627-20-3	2-Pentene, (Z)-	Liquid	44,520
463-49-0	Propadiene [1,2-Propadiene]	Gas	46,332
74-98-6	Propane	Gas	46,333
115-07-1	Propylene [1-Propene]	Gas	45,762
74-99-7	Propyne [1-Propyne]	Gas	46,165
7803-62-5	Silane	Gas	44,307

Exhibit C-1 (continued)

CAS No.	Chemical Name	Physical State at 25° C	Heat of Combustion (kjoule/kg)
116-14-3	Tetrafluoroethylene [Ethene, tetrafluoro-]	Gas	1,284
75-76-3	Tetramethylsilane [Silane, tetramethyl-]	Liquid	41,712
10025-78-2	Trichlorosilane [Silane, trichloro-]	Liquid	3,754
79-38-9	Trifluorochloroethylene [Ethene, chlorotrifluoro-]	Gas	1,837
75-50-3	Trimethylamine [Methanamine, N,N-dimethyl-]	Gas	37,978
689-97-4	Vinyl acetylene [1-Buten-3-yne]	Gas	45,357
75-01-4	Vinyl chloride [Ethene, chloro-]	Gas	18,848
109-92-2	Vinyl ethyl ether [Ethene, ethoxy-]	Liquid	32,909
75-02-5	Vinyl fluoride [Ethene, fluoro-]	Gas	2,195
75-35-4	Vinylidene chloride [Ethene, 1,1-dichloro-]	Liquid	10,354
75-38-7	Vinylidene fluoride [Ethene, 1,1-difluoro-]	Gas	10,807
107-25-5	Vinyl methyl ether [Ethene, methoxy-]	Gas	30,549

* Estimated heat of combustion

Exhibit C-2
Data for Flammable Gases

CAS Number	Chemical Name	Molecular Weight	Ratio of Specific Heats	Flammability Limits (Vol %)		LFL (mg/L)	Gas Factor (GF)[g]	Liquid Factor Boiling (LFB)	Density Factor (Boiling) (DF)	Reference Table[a]	Pool Fire Factor (PFF)	Flash Fraction Factor (FFF)[f]
				Lower (LFL)	Upper UFL							
75-07-0	Acetaldehyde	44.05	1.18	4.0	60.0	72	22	0.11	0.62	Dense	2.7	0.018
74-86-2	Acetylene	26.04	1.23	2.5	80.0	27	17	0.12	0.78	Buoyant[b]	4.8	0.23[f]
598-73-2	Bromotrifluoroethylene	160.92	1.11	[c]	37.0	[c]	41[c]	0.25[c]	0.29[c]	Dense	0.42[c]	0.15[c]
106-99-0	1,3-Butadiene	54.09	1.12	2.0	11.5	44	24	0.14	0.75	Dense	5.5	0.15
106-97-8	Butane	58.12	1.09	1.5	9.0	36	25	0.14	0.81	Dense	5.9	0.15
25167-67-3	Butene	56.11	1.10	1.7	9.5	39	24	0.14	0.77	Dense	5.6	0.14
590-18-1	2-Butene-cis	56.11	1.12	1.6	9.7	37	24	0.14	0.76	Dense	5.6	0.11
624-64-6	2-Butene-trans	56.11	1.11	1.8	9.7	41	24	0.14	0.77	Dense	5.6	0.12
106-98-9	1-Butene	56.11	1.11	1.6	9.3	37	24	0.14	0.78	Dense	5.7	0.17
107-01-7	2-Butene	56.11	1.10	1.7	9.7	39	24	0.14	0.77	Dense	5.6	0.12
463-58-1	Carbon oxysulfide	60.08	1.25	12.0	29.0	290	26	0.18	0.41	Dense	1.3	0.29
7791-21-1	Chlorine monoxide	86.91	1.21	23.5	NA	830	31	0.19	NA	Dense	0.15	NA
557-98-2	2-Chloropropylene	76.53	1.12	4.5	16.0	140	29	0.16	0.54	Dense	3.3	0.011
460-19-5	Cyanogen	52.04	1.17	6.0	32.0	130	24	0.15	0.51	Dense	2.5	0.40
75-19-4	Cyclopropane	42.08	1.18	2.4	10.4	41	22	0.13	0.72	Dense	5.4	0.23
4109-96-0	Dichlorosilane	101.01	1.16	4.0	96.0	160	33	0.20	0.40	Dense	1.3	0.084
75-37-6	Difluoroethane	66.05	1.14	3.7	18.0	100	27	0.17	0.48	Dense	1.6	0.23
124-40-3	Dimethylamine	45.08	1.14	2.8	14.4	52	22	0.12	0.73	Dense	3.7	0.090
463-82-1	2,2-Dimethylpropane	72.15	1.07	1.4	7.5	41	27	0.16	0.80	Dense	6.4	0.11
74-84-0	Ethane	30.07	1.19	2.9	13.0	36	18	0.14	0.89	Dense	5.4	0.75
107-00-6	Ethyl acetylene	54.09	1.11	2.0	32.9	44	24	0.13	0.73	Dense	5.4	0.091
75-04-7	Ethylamine	45.08	1.13	3.5	14.0	64	22	0.12	0.71	Dense	3.6	0.040

Exhibit C-2 (continued)

CAS Number	Chemical Name	Molecular Weight	Ratio of Specific Heats	Flammability Limits (Vol %)		LFL (mg/L)	Gas Factor (GF)[g]	Liquid Factor Boiling (LFB)	Density Factor (Boiling) (DF)	Reference Table[a]	Pool Fire Factor (PFF)	Flash Fraction Factor (FFF)[f]
				Lower (LFL)	Upper UFL							
75-00-3	Ethyl chloride	64.51	1.15	3.8	15.4	100	27	0.15	0.53	Dense	2.6	0.053
74-85-1	Ethylene	28.05	1.24	2.7	36.0	31	18	0.14	0.85	Buoyant[b]	5.4	0.63[f]
109-95-5	Ethyl nitrite	75.07	1.30	4.0	50.0	120	30	0.16	0.54	Dense	2.0	NA
1333-74-0	Hydrogen	2.02	1.41	4.0	75.0	3.3	5.0	[c]	[c]	[d]	[c]	NA
75-28-5	Isobutane	58.12	1.09	1.8	8.4	43	25	0.15	0.82	Dense	6.0	0.23
74-82-8	Methane	16.04	1.30	5.0	15.0	33	14	0.15	1.1	Buoyant	5.6	0.87[f]
74-89-5	Methylamine	31.06	1.19	4.9	20.7	62	19	0.10	0.70	Dense	2.7	0.12
563-45-1	3-Methyl-1-butene	70.13	1.08	1.5	9.1	43	26	0.15	0.77	Dense	6.0	0.030
115-10-6	Methyl ether	46.07	1.15	3.3	27.3	64	22	0.14	0.66	Dense	3.4	0.22
115-11-7	2-Methylpropene	56.11	1.10	1.8	8.8	41	24	0.14	0.77	Dense	5.7	0.18
463-49-0	Propadiene	40.07	1.16	2.1	2.1	34	21	0.13	0.73	Dense	5.2	0.20
74-98-6	Propane	44.10	1.13	2.0	9.5	36	22	0.14	0.83	Dense	5.7	0.38
115-07-1	Propylene	42.08	1.15	2.0	11.0	34	21	0.14	0.79	Dense	5.5	0.35
74-99-7	Propyne	40.07	1.16	1.7	39.9	28	21	0.12	0.72	Dense	4.9	0.18
7803-62-5	Silane	32.12	1.24	[c]	[c]	[c]	19 [c]	[c]	[c]	Dense	[c]	0.41[f]
116-14-3	Tetrafluoroethylene	100.02	1.12	11.0	60.0	450	33	0.29	0.32	Dense	0.25	0.69
79-38-9	Trifluorochloroethylene	116.47	1.11	8.4	38.7	400	35	0.26	0.33	Dense	0.34	0.27
75-50-3	Trimethylamine	59.11	1.10	2.0	11.6	48	25	0.14	0.74	Dense	4.8	0.12
689-97-4	Vinyl acetylene	52.08	1.13	2.2	31.7	47	24	0.13	0.69	Dense	5.4	0.086
75-01-4	Vinyl chloride	62.50	1.18	3.6	33.0	92	26	0.16	0.50	Dense	2.4	0.14
75-02-5	Vinyl fluoride	46.04	1.20	2.6	21.7	49	23	0.17	0.57	Dense	0.28	0.37
75-38-7	Vinylidene fluoride	64.04	1.16	5.5	21.3	140	27	0.22	0.42	Dense	1.8	0.50

Exhibit C-2 (continued)

CAS Number	Chemical Name	Molecular Weight	Ratio of Specific Heats	Flammability Limits (Vol %)		LFL (mg/L)	Gas Factor (GF)[g]	Liquid Factor Boiling (LFB)	Density Factor (Boiling) (DF)	Reference Table[a]	Pool Fire Factor (PFF)	Flash Fraction Factor (FFF)[f]
				Lower (LFL)	Upper UFL							
107-25-5	Vinyl methyl ether	58.08	1.12	2.6	39.0	62	25	0.17	0.57	Dense	3.7	0.093

Notes:

NA: Data not available

[a] "Buoyant" in the Reference Table column refers to the tables for neutrally buoyant gases and vapors; "Dense" refers to the tables for dense gases and vapors. See Appendix D, Section D.4.4, for more information on the choice of reference tables.

[b] Gases that are lighter than air may behave as dense gases upon release if liquefied under pressure or cold; consider the conditions of release when choosing the appropriate table.

[c] Reported to be spontaneously combustible.

[d] Much lighter than air; table of distances for neutrally buoyant gases not appropriate.

[e] Pool formation unlikely.

[f] Calculated at 298 K (25 °C) with the following exceptions:

Acetylene factor at 250 K as reported in TNO, *Methods for the Calculation of the Physical Effects of the Escape of Dangerous Material* (1980).

Ethylene factor calculated at critical temperature, 282 K.

Methane factor calculated at critical temperature, 191 K.

Silane factor calculated at critical temperature, 270 K.

[g] Use GF for gas leaks under choked (maximum) flow conditions.

Exhibit C-3
Data for Flammable Liquids

CAS Number	Chemical Name	Molecular Weight	Flammability Limit (Vol%)		LFL (mg/L)	Liquid Factors		Density Factor	Liquid Leak Factor (LLF)[a]	Reference Table[b]	Pool Fire Factor (PFF)
			Lower (LFL)	Upper (UFL)		Ambient (LFA)	Boiling (LFB)				
590-21-6	1-Chloropropylene	76.53	4.5	16.0	140	0.11	0.15	0.52	45	Dense	3.2
60-29-7	Ethyl ether	74.12	1.9	48.0	57	0.11	0.15	0.69	34	Dense	4.3
75-08-1	Ethyl mercaptan	62.14	2.8	18.0	71	0.10	0.13	0.58	40	Dense	3.3
78-78-4	Isopentane	72.15	1.4	7.6	41	0.14	0.15	0.79	30	Dense	6.1
78-79-5	Isoprene	68.12	2.0	9.0	56	0.11	0.14	0.72	32	Dense	5.5
75-31-0	Isopropylamine	59.11	2.0	10.4	48	0.10	0.13	0.71	33	Dense	4.1
75-29-6	Isopropyl chloride	78.54	2.8	10.7	90	0.11	0.16	0.57	41	Dense	3.1
563-46-2	2-Methyl-1-butene	70.13	1.4	9.6	40	0.12	0.15	0.75	31	Dense	5.8
107-31-3	Methyl formate	60.05	5.9	20.0	140	0.10	0.13	0.50	46	Dense	1.8
504-60-9	1,3-Pentadiene	68.12	1.6	13.1	44	0.077	0.14	0.72	33	Dense	5.3
109-66-0	Pentane	72.15	1.3	8.0	38	0.10	0.15	0.78	30	Dense	5.8
109-67-1	1-Pentene	70.13	1.5	8.7	43	0.13	0.15	0.77	31	Dense	5.8
646-04-8	2-Pentene, (E)-	70.13	1.4	10.6	40	0.10	0.15	0.76	31	Dense	5.6
627-20-3	2-Pentene, (Z)-	70.13	1.4	10.6	40	0.10	0.15	0.75	31	Dense	5.6
75-76-3	Tetramethylsilane	88.23	1.5	NA	54	0.17	0.17	0.59	40	Dense	6.3
10025-78-2	Trichlorosilane	135.45	1.2	90.5	66	0.18	0.23	0.37	64	Dense	0.68
109-92-2	Vinyl ethyl ether	72.11	1.7	28.0	50	0.10	0.15	0.65	36	Dense	4.2
75-35-4	Vinylidene chloride	96.94	7.3	NA	290	0.15	0.18	0.44	54	Dense	1.6

Notes:

NA: Data not available.
[a] Use the LLF only for leaks from tanks at atmospheric pressure.
[b] "Dense" in the Reference Table column refers to the tables for dense gases and vapors. See Appendix D, Section D.4.4, for more information on the choice of reference tables.

APPENDIX D

TECHNICAL BACKGROUND

APPENDIX D TECHNICAL BACKGROUND

D.1 Worst-Case Release Rate for Gases

D.1.1 Unmitigated Release

The assumption that the total quantity of toxic gas is released in 10 minutes is the same assumption used in EPA's *Technical Guidance for Hazards Analysis* (1987).

D.1.2 Gaseous Release Inside Building

The mitigation factor for gaseous release inside a building is based on a document entitled, *Risk Mitigation in Land Use Planning: Indoor Releases of Toxic Gases*, by S.R. Porter. This paper presented three release scenarios and discussed the mitigating effects that would occur in a building with a volume of 1,000 cubic meters at three different building air exchange rates. There is a concern that a building may not be able to withstand the pressures of a very large release. However, this paper indicated that release rates of at least 2,000 pounds per minute could be withstood by a building.

Analyzing the data in this paper several ways, the value of 55 percent emerged as representing the mitigation that could occur for a release scenario into a building. Data are provided on the maximum release rate in a building and the maximum release rate from a building. Making this direct comparison at the lower maximum release rate (3.36 kg/s) gave a release rate from the building of 55 percent of the release rate into the building. Using information provided on another maximum release rate (10.9 kg/min) and accounting for the time for the release to accumulate in the building, approximately 55 percent emerged again.

The choice of building ventilation rates affects the results. The paper presented mitigation for three different ventilation rates, 0.5, 3, and 10 air changes per hour. A ventilation rate of 0.5 changes per hour is representative of specially designed, "gas-tight" buildings, based on the Porter reference. EPA decided that this ventilation rate was appropriate for this analysis. A mitigation factor of 55 percent may be used in the event of a gaseous release which does not destroy the building into which it is released. This factor may overstate the mitigation provided by a building with a higher ventilation rate.

For releases of ammonia, chlorine, and sulfur dioxide, factors specific to the chemicals, the conditions of the release, and building ventilation rates have been developed to estimate mitigation of releases in buildings. For information on these factors and estimation of mitigated release rates, see *Backup Information for the Hazard Assessments in the RMP Offsite Consequence Analysis Guidance, the Guidance for Wastewater Treatment Facilities and the Guidance for Ammonia Refrigeration - Anhydrous Ammonia, Aqueous Ammonia, Chlorine and Sulfur Dioxide*. See also the industry-specific guidance documents for ammonia refrigeration and POTWs.

D.2 Worst-Case Release Rate for Liquids

D.2.1 Evaporation Rate Equation

The equation for estimating the evaporation rate of a liquid from a pool is from the *Technical Guidance for Hazards Analysis*, Appendix G. The same assumptions are made for determination of maximum pool area (i.e., the pool is assumed to be 1 centimeter (0.033 feet) deep). The evaporation rate equation has been modified to include a different mass transfer coefficient for water, the reference compound. For this document, a value of 0.67 centimeters per second is used as the mass transfer coefficient, instead of the value of 0.24 cited in the *Technical Guidance for Hazards Analysis*. The value of 0.67 is based on Donald MacKay and Ronald S. Matsugu, "Evaporation Rates of Liquid Hydrocarbon Spills on Land and Water," *Canadian Journal of Chemical Engineering*, August 1973, p. 434. The evaporation equation becomes:

$$QR = \frac{0.284 \times U^{0.78} \times MW^{2/3} \times A \times VP}{82.05 \times T} \qquad \textbf{(D-1)}$$

where:

QR	=	Evaporation rate (pounds per minute)
U	=	Wind speed (meters per second)
MW	=	Molecular weight (given in Exhibits B-1 and B-2, Appendix B, for toxic substances and Exhibits C-2 and C-3, Appendix C, for flammable substances)
A	=	Surface area of pool formed by the entire quantity of the mixture (square feet) (determined as described in Section 3.2.2 of the text)
VP	=	Vapor pressure (mm Hg)
T	=	Temperature of released substance (Kelvin (K); temperature in °C plus 273, or 298 for 25 °C)

D.2.2 Factors for Evaporation Rate Estimates

Liquid Factors. The liquid factors, Liquid Factor Ambient (LFA) and Liquid Factor Boiling (LFB), used to estimate the evaporation rate from a liquid pool (see Section 3.2 of this guidance document), are derived as described in the *Technical Guidance for Hazards Analysis*, Appendix G, with the following differences:

- The mass transfer coefficient of water is assumed to be 0.67, as discussed above; the value of the factor that includes conversion factors, the mass transfer coefficient for water, and the molecular weight of water to the one-third power, given as 0.106 in the *Technical Guidance* is 0.284 in this guidance.

- Density of all substances was assumed to be the density of water in the *Technical Guidance*; the density was included in the liquid factors. For this guidance document, density is not included in the LFA and LFB values presented in the tables; instead, a separate Density Factor (DF) (discussed below) is provided to be used in the evaporation rate estimation.

With these modifications, the LFA is:

$$LFA = \frac{0.284 \times MW^{2/3} \times VP}{82.05 \times 298} \qquad \textbf{(D-2)}$$

where: MW = Molecular weight

VP	=	Vapor pressure at ambient temperature (mm Hg)
298 K (25 °C)	=	Ambient temperature and temperature of released substance

LFB is:

$$LFB = \frac{0.284 \times MW^{2/3} \times 760}{82.05 \times BP} \tag{D-3}$$

where:			
	MW	=	Molecular weight
	760	=	Vapor pressure at boiling temperature (mm Hg)
	BP	=	Boiling point (K)

LFA and LFB values were developed for all toxic and flammable regulated liquids, and LFB values, to be used for analysis of gases liquefied by refrigeration, were developed for toxic and flammable gases.

Density Factor. Because some of the regulated liquids have densities very different from that of water, the density of each substance was used to develop a Density Factor (DF) for the determination of maximum pool area for the evaporation rate estimation. DF values were developed for toxic and flammable liquids at ambient temperature and for toxic and flammable gases at their boiling points. The density factor is:

$$DF = \frac{1}{d \times 0.033} \tag{D-4}$$

where:			
	DF	=	Density factor ($1/(lbs/ft^2)$)
	d	=	Density of the substance in pounds per cubic foot
	0.033	=	Depth of pool for maximum area (feet)

Temperature Correction Factors. Temperature correction factors were developed for toxic liquids released at temperatures above 25 °C, the temperature used for development of the LFAs. The temperature correction factors are based on vapor pressures calculated from the coefficients provided in *Physical and Thermodynamic Properties of Pure Chemicals, Data Compilation*, developed by the Design Institute for Physical Property Data (DIPPR), American Institute of Chemical Engineers. The factors are calculated as follows:

$$TCF_T = \frac{VP_T \times 298}{VP_{298} \times T} \tag{D-5}$$

where:			
	TCF_T	=	Temperature Correction Factor at temperature T
	VP_T	=	Vapor pressure at temperature T
	VP_{298}	=	Vapor pressure at 298 K
	T	=	Temperature (K) of released substance

Factors were developed at intervals of 5 °C for temperatures up to 50 °C.

No correction factor was deemed necessary for changes in the density of the regulated toxic liquids with changes in temperature, although the density could affect the pool area and release rate estimates. Analysis of the temperature dependence of the density of these liquids indicated that the changes in density with temperature were very small compared to the changes in vapor pressure with temperature.

D.2.3 Common Water Solutions and Oleum

Water solutions of regulated toxic substances must be analyzed somewhat differently from pure toxic liquids. Except for solutions of relatively low concentration, the evaporation rate varies with the concentration of the solution. At one specific concentration, the composition of the liquid does not change as evaporation occurs. For concentrated solutions of volatile substances, the evaporation rate from a pool may decrease, very rapidly in some cases, as the toxic substance volatilizes and its concentration in the pool decreases. To analyze these changes, EPA used spreadsheets to estimate the vapor pressure, concentration, and release rate at various time intervals for regulated toxic substances in water solution evaporating from pools. In addition to the spreadsheet analysis, EPA used the ALOHA model with an additional step-function feature (not available in the public version). With this step-function feature, changes in the release rate could be incorporated and the effects of these changes on the consequence distance analyzed. The results of the spreadsheet calculations and the model were found to be in good agreement. The distance results obtained from the spreadsheet analysis and the model for various solutions were compared with the results from various time averages to examine the sensitivity of the results. An averaging time of 10 minutes was found to give reasonable agreement with the step-function model for most substances at various concentrations. The spreadsheet analysis also indicated that the first 10 minutes of evaporation was the most important, and the evaporation rate in the first 10 minutes likely could be used to estimate the distance to the endpoint.

Oleum is a solution of sulfur trioxide in sulfuric acid. Sulfur trioxide evaporating from oleum exhibits release characteristics similar to those of toxic substances evaporating from water solutions. Analysis of oleum releases, therefore, was carried out in the same way as for water solutions.

NOAA developed a computerized calculation method to estimate partial vapor pressures and release rates for regulated toxic substance in solution as a function of concentration, based on vapor pressure data from *Perry's Chemical Engineers' Handbook* and other sources. Using this method and spreadsheet calculations, EPA estimated partial vapor pressures and evaporation rates at one-minute intervals over 10 minutes for solutions of various concentrations. The 10-minute time period was chosen based on the ALOHA results and other calculations. For each one-minute interval, EPA estimated the concentration of the solution based on the quantity evaporated in the previous interval and estimated the partial vapor pressure based on the concentration. These estimated vapor pressures were used to calculate an average vapor pressure over the 10-minute period; this average vapor pressure was used to derive Liquid Factor Ambient (LFA) values, as described above for liquids. Use of these factors is intended to give an evaporation rate that accounts for the decrease in evaporation rate expected to take place as the solution evaporates.

Density Factors (DF) were developed for solutions of various concentrations from data in *Perry's Chemical Engineers' Handbook* and other sources, as discussed above for liquids.

Because solutions do not have defined boiling points, EPA did not develop Liquid Factor Boiling (LFB) values for solutions. As a simple and conservative approach, the quantity of a regulated substance in a solution at an elevated temperatures is treated as a pure substance. The LFB for the pure substance, or the

LFA and a temperature correction factor, is used to estimate the initial evaporation rate of the regulated substance from the solution. Only the first 10 minutes of evaporation are considered, as for solutions at ambient temperatures, because the release rate would decrease rapidly as the substance evaporates and the concentration in the solution decreases. This approach will likely give an overestimate of the release rate and of the consequence distance.

D.2.4 Releases Inside Buildings

If a liquid is released inside a building, its release to the outside air will be mitigated in two ways. First, the evaporation rate of the liquid may be much lower inside a building than outside. This is due to wind speed, which directly affects the evaporation rate. The second mitigating factor is that the building provides resistance to discharge of contaminated air to the outdoors.

In this method, a conservative wind speed, U, of 0.1 meter per second (m/s) was assumed in the building. (See end of text for a justification of this wind speed.) For a release outdoors in a worst-case scenario, U is set to 1.5 m/s, and for an alternative scenario, U is set to 3 m/s. The evaporation rate equation is:

$$QR = U^{0.78} \times (LFA,\ LFB) \times A \qquad \textbf{(D-6)}$$

where:

where:	QR	=	Release rate (pounds per minute (lbs/min))
	U	=	Wind speed (meters per second (m/s))
	LFA	=	Liquid Factor Ambient
	LFB	=	Liquid Factor Boiling
	A	=	Area of pool (square feet (ft^2))

As can be seen, if U inside a building is only 0.1, then the evaporation rate inside a building will be much lower than a corresponding evaporation rate outside (assuming the temperature is the same). The rate will only be $(0.1/1.5)^{0.78}$, about 12 percent of the rate for a worst case, and $(0.1/3)^{0.78}$, about seven percent of the rate for an alternative case.

The evaporated liquid mixes with and contaminates the air in the building. What EPA is ultimately interested in is the rate at which this contaminated air exits the building. In order to calculate the release of contaminated air outside the building, EPA adapted a method from an UK Health and Safety Executive paper entitled, *Risk Mitigation in Land Use Planning: Indoor Releases of Toxic Gases*, by S.R. Porter. EPA assumed that the time for complete evaporation of the liquid pool was one hour. The rate at which contaminated air was released from the building during liquid evaporation (based on the paper) was assumed to be equal to the evaporation rate plus the building ventilation rate (no pressure buildup in building). The building ventilation rate was set equal to 0.5 air changes per hour. This ventilation rate is representative of a specially designed, "gas-tight" building. (The mitigation factor developed based on this type of building would overstate the mitigation provided by a building with higher ventilation rates.) EPA used a building with a volume of 1,000 cubic meters (m^3) and a floor area of 200 m^2 (2,152 ft^2) as an example for this analysis. EPA assumed that the liquid pool would cover the entire building floor, representing a conservative scenario.

To provide a conservative estimate, EPA calculated the evaporation rate for a spill of a volatile liquid, carbon disulfide (CS_2), under ambient conditions inside the building:

$$QR = 0.1^{0.78} \times 0.075 \times 2{,}152 = 26.8 \text{ pounds per minute (lbs/min)}$$

Next, this evaporation rate was converted to cubic meters per minute (m^3/min) using the ideal gas law (the molecular weight of CS_2 is 76.1):

26.8 lbs/min × 454 grams per pound (g/lb) × 1 mole CS_2/76.1 g × 0.0224 m^3/mole = 3.58 m^3/min.

The ventilation rate of the building is 0.5 changes per hour, which equals 500 m^3 per hour, or 8.33 m^3/min. Therefore, during evaporation, contaminated air is leaving the building at a rate of 8.33 + 3.58, or 11.9 m^3/min.

EPA used an iterative calculation for carbon disulfide leaving a building using the above calculated parameters. During the first minute of evaporation, 26.8 lbs of pure carbon disulfide evaporates, and EPA assumed this evenly disperses through the building so that the concentration of CS_2 in the building air is 0.0268 lbs/m^3 (assuming 1000 m^3 volume in the building). Contaminated air is exiting the building at a rate of 11.9 m^3/min, so EPA deduced that 11.9 × 0.0268 = 0.319 lbs of carbon disulfide exit the building in the first minute, leaving 26.5 lbs still evenly dispersed inside. Since this release occurs over one minute, the release rate of the carbon disulfide to the outside is 0.319 lbs/min. During the second minute, another 26.8 lbs of pure carbon disulfide evaporates and disperses, so that the building now contains 26.8 + 26.5 = 53.3 lbs of carbon disulfide, or 0.0533 lbs/m^3. Contaminated air is still exiting the building at a rate of 11.9 m^3/min, so 11.9 × 0.05328 = 0.634 lbs of carbon disulfide are released, leaving 52.6 lbs inside. Again, this release occurs over one minute so that the rate of carbon disulfide exiting the building in terms of contaminated air is 0.634 lbs/min. EPA continued to perform this estimation over a period of one hour. The rate of release of carbon disulfide exiting the building in the contaminated air at the sixty minute mark is 13.7 lbs/min. This represents the maximum rate of carbon disulfide leaving the building. After all of the carbon disulfide is evaporated, there is a drop in the concentration of carbon disulfide in the contaminated air leaving the building because the evaporation of carbon disulfide no longer contributes to the overall contamination of the air.

Note that if the same size pool of carbon disulfide formed outside, the release rate for a worst-case scenario would be:

$$QR = 1.5^{0.78} \times 0.075 \times 2{,}152 = 221 \text{ lbs/min}$$

and for an alternative case:

$$QR = 3^{0.78} \times 0.075 \times 2{,}152 = 380 \text{ lbs/min.}$$

The maximum release rate of carbon disulfide in the contaminated building air, assuming a 1,000 m^3 building with a building exchange rate of 0.5 air changes per hour, was only about 6 percent (13.7 ÷ 221 lbs/min x 100) of the worst-case scenario rate, and only about 3.6 percent (13.7 ÷ 380 lbs/min x 100) of the alternative scenario rate. EPA set an overall building mitigation factor equal to 10 percent and five percent, respectively, in order to be conservative. Please note that (at a constant ventilation rate of 0.5 changes per hour) as the size of the building increases, the maximum rate of contaminated air leaving the building will decrease, although only slightly, because of the balancing effect of building volume and ventilation rate. Obviously, a higher ventilation rate will yield a higher maximum release rate of contaminated air from the building.

For a release inside a building, EPA assumed a building air velocity of 0.1 m/s. This conservative value was derived by setting the size of the ventilation fan equal to 1.0 m^2. This fan is exchanging air from the building with the outside at a rate of 0.5 changes per hour. For a 1,000 m^3 building, this value becomes 500 m^3/hour, or 0.14 m^3/s. Dividing 0.14 m^3/s by the area of the fan yields a velocity of 0.14 m/s, which was rounded down to 0.1 m/s.

D.3 Toxic Endpoints

The toxic endpoints for regulated toxic substances, which are specified in the RMP Rule, are presented in Appendix B, Exhibits B-1, B-2, and B-3. The endpoints were chosen as follows, in order of preference:

(1) Emergency Response Planning Guideline 2 (ERPG-2), developed by the American Industrial Hygiene Association, if available;

(2) Level of Concern (LOC) derived for extremely hazardous substances (EHSs) regulated under section 302 of the Emergency Planning and Community Right-to-Know Act (EPCRA) (see the *Technical Guidance for Hazards Analysis* for more information on LOCs); the LOC for EHSs is based on:

-- One-tenth of the Immediately Dangerous to Life and Health (IDLH) level, developed by the National Institute of Occupational Safety and Health (NIOSH), using IDLH values developed before 1994,

or, if no IDLH value is available,

-- One-tenth of an estimated IDLH derived from toxicity data; the IDLH is estimated as described in Appendix D of the *Technical Guidance for Hazards Analysis*.

Note that the LOCs were not updated using IDLHs published in 1994 and later, because NIOSH revised its methodology for the IDLHs. The EHS LOCs based on earlier IDLHs were reviewed by EPA's Science Advisory Board, and EPA decided to retain the methodology that was reviewed.

ERPG-2 is defined as the maximum airborne concentration below which it is believed nearly all individuals could be exposed for up to one hour without experiencing or developing irreversible or other serious health effects or symptoms that could impair an individual's ability to take protective action.

IDLH (pre-1994) concentrations were defined in the NIOSH *Pocket Guide to Chemical Hazards* as representing the maximum concentration from which, in the event of respirator failure, one could escape within 30 minutes without a respirator and without experiencing any escape-impairing (e.g., severe eye irritation) or irreversible health effects. (As noted above, LOCs for EHSs were not updated to reflect 1994 and later IDLHs.)

The estimated IDLH is derived from animal toxicity data, in order of preferred data, as follows:

- From median lethal concentration (LC_{50}) (inhalation): 0.1 x LC_{50}
- From lowest lethal concentration (LC_{LO}) (inhalation): 1 x LC_{LO}
- From median lethal dose (LD_{50}) (oral): 0.01 x LD_{50}
- From lowest lethal dose (LD_{LO}) (oral): 0.1 x LD_{LO}

The toxic endpoints based on LOCs for EHSs presented in the tables in Appendix B are, in some cases, different from the LOCs listed in the *Technical Guidance for Hazards Analysis*, because some of the LOCs were updated based on IDLHs that were published after the development of the LOCs (and before 1994) or on new or revised toxicity data.

D.4 Reference Tables for Distances to Toxic and Flammable Endpoints

D.4.1 Neutrally Buoyant Gases

Toxic Substances. Reference tables for distances to toxic endpoints for neutrally buoyant gases and vapors were derived from the Gaussian model using the longitudinal dispersion coefficients based on work by Beals (*Guide to Local Diffusion of Air Pollutants*, Technical Report 214. Scott Air Force Base, Illinois: U.S. Air Force, Air Weather Service, 1971). The reasons for using the Beals dispersion coefficients are discussed below.

Longitudinal dispersion (dispersion in the along-wind direction) is generated mostly by vertical wind shear. Wind shear results from the tendency of the wind speed to assume a wind profile—the speed is lowest next to the ground and increases with height until it reaches an asymptotic value at approximately a few hundred feet above the surface. To account for shear-driven dispersion, any air dispersion model intended for modeling short-duration releases must include either (a) a formulation that accounts, either implicitly or explicitly, for the height-dependence of wind speed or (b) some type of parameterization that converts shear effect into σ_x, the standard deviation function in the along-wind direction.

Because the standard Gaussian formula does not incorporate σ_x (it includes only σ_y and σ_z, the crosswind and horizontal functions), very few alternate ways to formulate σ_x have been proposed. The simplest method was proposed by Turner (*Workbook of Atmospheric Dispersion Estimates*, Report PB-191 482. Research Triangle Park, North Carolina: Office of Air Programs, U.S. Environmental Protection Agency, 1970), who suggested simply setting σ_x equal to σ_y. Textbooks such as that by Pasquill and Smith (*Atmospheric Diffusion*, 3rd ed. New York: Halstead Press, 1983) describe a well-known analytic model. However, this model is more complex than a Gaussian model because according to it, dispersion depends on wind shear and the vertical variation of the vertical diffusion coefficient. Wilson (Along-wind Diffusion of Source Transients, *Atmospheric Environment* 15:489-495, 1981) proposed another method in which σ_x is determined as a function of wind shear, but in a form that can then be used in a Gaussian model. However, it is now believed that Wilson's formulation gives σ_x values that are too large.

To avoid the problems of the analytic method and Wilson's formulation, we chose to include a formulation for σ_x derived from work by Beals (1971). We had three reasons for doing so. First, in terms of magnitude, Beals' σ_x fell in the midrange of the alternative formulations that we reviewed. Second, Beals' σ_x indirectly accounts for wind shear by using (unpublished) experimental data. Third, both the ALOHA and DEGADIS models incorporate the Beals methodology.

When a substance is dispersed downwind, the concentration in the air changes over time. To assess the health effects of potential exposure to the substance, the average concentration of the substance over some time period is determined. Averaging time is the time interval over which the instantaneous concentration of the hazardous material in the vapor cloud is averaged. Averaging time should generally be equal to or shorter than either the release duration or cloud duration and, if possible, should reflect the exposure time associated with the toxic exposure guideline of interest. The exposure time associated with the toxic endpoints specified under the RMP Rule include 30 minutes for the Immediately Dangerous to Life and Health (IDLH) level and 60 minutes for the Emergency Response Planning Guideline (ERPG). For the neutrally buoyant tables, the 10-minute release scenario was modeled using a 10-minute averaging time. The 60-minute release scenario was modeled using a 30-minute averaging time to be consistent with the 30-minute exposure time associated with the IDLH. A 60-minute averaging time may have underpredicted consequence distances and, therefore, was not used for development of the distance tables for this guidance.

Cloud dispersion from a release of finite duration (10 and 60-minute releases) is calculated using an equation specified in the NOAA publication *ALOHA™ 5.0 Theoretical Description*, Technical Memorandum NOS ORCA 65, August 1992.

Flammable Substances. The reference tables of distances for vapor cloud fires of neutrally buoyant flammable substances were derived using the same model as for toxic substances, as described above. The endpoint for modeling was the lower flammability limit (LFL). For flammable substances, an averaging time of 0.1 minute (six seconds) was used, because fires are considered to be nearly instantaneous events.

Distances of interest for flammable substances are generally much shorter than for toxic substance, because the LFL concentrations are much larger than the toxic endpoints. For the short distances found in modeling the flammable substances, modeling results were found to be the same for 10-minute and longer releases; therefore, one table of distances for rural conditions and one table for urban conditions, applicable for both 10-minute and longer releases, were developed for flammable substances.

D.4.2 Dense Gases

Toxic Substances. The reference tables for dense gases were developed using the widely accepted SLAB model, developed by Lawrence Livermore National Laboratory. SLAB solves conservation equations of mass, momentum, energy, and species for continuous, finite duration, and instantaneous releases. The reference tables were based on the evaporating pool algorithm.

For the reference tables were developed based on modeling releases of hydrogen chloride (HCl). HCl was chosen based on a SLAB modeling analysis of a range of dispersion behavior for releases of regulated dense gases or vapors with different molecular weights. This analysis showed that releases of HCl generally provided conservative results under a variety of stability/wind speed combinations, release rates, and toxic endpoints.

Similar to the modeling of neutrally buoyant plumes, the 10-minute release scenario of toxic chemicals was modeled using a 10-minute averaging time. The 60-minute release scenario was modeled using a 30-minute averaging time to be consistent with the 30-minute exposure time associated with the IDLH.

For all dense gas tables, the reference height for the wind speed was 10 meters. Relative humidity was assumed to be 50 percent, and the ambient temperature was 25 °C. The source area was the smallest value that still enabled the model to run for all release rates. The surface roughness factor was one meter for urban scenarios and three centimeters for rural scenarios.

Flammable Substances. For the reference tables for dispersion of dense flammable gases and vapors, for analysis of vapor cloud fires, the same model was used as for toxic substances, as described above, and the same assumptions were made. For the dispersion of flammable chemicals, averaging time should be very small (i.e., no more than a few seconds), because flammable vapors need only be exposed to an ignition source for a short period of time to initiate the combustion process. Thus, both the 10-minute and 60-minute reference tables for flammable substances use an averaging time of 10 seconds. The 10-minute and 60-minute tables were combined for flammable substances because the modeling results were found to be the same.

D.4.3 Chemical-Specific Reference Tables

The chemical-specific reference tables of distances for ammonia, chlorine, and sulfur dioxide were developed for EPA's risk management program guidance for ammonia refrigeration and for POTWs. For information on the chemical-specific modeling and development of the chemical-specific reference tables, see *Backup Information for the Hazard Assessments in the RMP Offsite Consequence Analysis Guidance, the Guidance for Wastewater Treatment Facilities and the Guidance for Ammonia Refrigeration - Anhydrous Ammonia, Aqueous Ammonia, Chlorine and Sulfur Dioxide*. See also the industry-specific guidance documents for ammonia refrigeration and POTWs.

The modeling carried out for aqueous ammonia also is applied in this guidance to ammonia released as a neutrally buoyant plume in other situations. The tables of distances derived from this modeling would apply to evaporation of ammonia from a water solution, evaporation of ammonia liquefied by refrigeration, or ammonia releases from the vapor space of a vessel, because the ammonia would behave as a neutrally buoyant plume (or possibly buoyant in some cases).

D.4.4 Choice of Reference Table for Dispersion Distances

Gases. Exhibit B-1 of Appendix B indicates whether the reference tables for neutrally buoyant or dense gases should be used for each of the regulated toxic gases. Exhibit C-2, Appendix C, provides this information for flammable gases. The choice of reference table presented in these exhibits is based on the molecular weight of the regulated substance compared to air; however, a number of factors that may cause a substance with a molecular weight similar to or smaller than the molecular weight of air to behave as a dense gas should be considered in selecting the appropriate table. For example, a cold gas may behave as a dense gas, even if it is lighter than air at ambient temperature. Gases liquefied under pressure may be released as a mixture of vapor and liquid droplets; because of presence of liquid mixed with the vapor, a gas that is lighter than air may behave as a dense gas in such a release. A gas that polymerizes or forms hydrogen bonds (e.g., hydrogen fluoride) also may behave as a dense gas.

Liquids and Solutions. Exhibits B-2 and B-3, Appendix B, and Exhibit C-3, Appendix C, indicate the reference table of distances to be used for each regulated liquid. The methodology presented in this guidance for consequence analysis for liquids and solutions assumes evaporation from a pool. All of the liquids regulated under CAA section 112(r) have molecular weights greater than the molecular weight of air; therefore, their vapor would be heavier than air. However, because the vapor from a pool will mix with air as it evaporates, the initial density of the vapor with respect to air may not in all cases indicate whether the vapor released from a pool should be modeled as a dense gas or a neutrally buoyant gas. If the rate of release from the pool is relatively low, the vapor-air mixture that is generated may be neutrally buoyant even if the vapor is denser than air, because the mixture may contain a relatively small fraction of the denser-than-air vapor; i.e., it may be mostly air. This may be the case particularly for some of the regulated toxic liquids with relatively low volatility. All of the regulated flammable substances have relatively high volatility; the reference tables for dense gases are assumed to be appropriate for analyzing dispersion of these flammable liquids.

To identify toxic liquids with molecular weight greater than air that might behave as neutrally buoyant gases when evaporating from a pool, EPA used the ALOHA model for pool evaporation of a number of substances with a range of molecular weights and vapor pressures. Modeling was carried out for F stability and wind speed 1.5 meters per second (worst-case conditions) and for D stability and wind speed 3.0 meters per second (alternative-case conditions). Pool spread to a depth of one centimeter was assumed. Additional modeling was carried out for comparison assuming different pool areas and depths. The molecular weight-vapor pressure combinations at which ALOHA used the neutrally buoyant gas model were used to develop the reference table choices given in Exhibit B-2 (for liquids) and B-3 (for solutions) in Appendix B. The neutrally buoyant tables should generally give reasonable results for pool evaporation under ambient conditions when indicated for liquids. At elevated temperatures, however, evaporation rates will be greater, and the dense gas tables should be used.

The liquids for which the neutrally buoyant table is identified for the worst case probably can be expected to behave as neutrally buoyant vapors when evaporating from pools under ambient conditions in most situations, but there may be cases when they exhibit dense gas behavior. Other liquids, for which the neutrally buoyant tables are not indicated for the worst case, might release neutrally buoyant vapors under some conditions (e.g., relatively small pools, temperature not much above 25 °C). Similarly, the liquids for which the neutrally buoyant tables are indicated as appropriate for alternative scenario analysis probably can be considered to behave as neutrally buoyant vapors under the alternative scenario conditions in most cases; however, there may be cases where they will behave as dense gases, and there may be other liquids that in some cases would exhibit neutrally buoyant behavior when evaporating. The reference table choices shown in Exhibit B-2 are intended to reflect the most likely behavior of the substances; they will not predict behavior of the listed substances evaporating under all conditions.

D.4.5 Additional Modeling for Comparison

Modeling was carried out for two worst-case examples and two alternative-case examples, using two different models, for comparison with the results obtained from the methods and distance tables in this guidance. This modeling is discussed below.

ALOHA Model. The Areal Locations of Hazardous Atmospheres (ALOHA) system was developed jointly by NOAA and EPA. ALOHA Version 5.2.1 was used for the comparison modeling. The parameters for ALOHA modeling were the same as specified in this guidance document for worst-case and alternative scenarios. The substances modeled are included in ALOHA's chemical database, so no chemical data were entered for modeling. For consistency with the methodology used to develop the reference tables of distances, a wind speed height of 10 meters was selected for ALOHA modeling.

For all of the substances modeled, the direct source model was chosen for ALOHA modeling, and the release rate estimated using the guidance methodology was entered as the release rate for ALOHA. ALOHA selected the dense gas model to estimate the distances to the endpoints in all cases.

WHAZAN Model. The World Bank Hazard Analysis (WHAZAN) system was developed by Technica International in collaboration with the World Bank. The 1988 version of WHAZAN was used for the comparison modeling. The parameters for atmospheric stability, wind speed, and ambient temperature and humidity were the same as specified in this guidance document. For surface roughness, WHAZAN requires entry of a "roughness parameter," rather than a height. Based on the discussion of this parameter in the WHAZAN Theory Manual, a roughness parameter of 0.07 (corresponding to flat land, few trees) was chosen as equivalent to the surface roughness of 3 centimeters used to represent rural topography in modeling to develop the distance tables for this guidance. A roughness parameter of 0.17 (for woods or rural area or industrial site) was chosen as equivalent to 1 meter, which was used to develop the urban distance tables. Data were added to the WHAZAN chemical database for acrylonitrile and allyl alcohol; ethylene oxide and chlorine were already included in the database.

For WHAZAN modeling of the gases ethylene oxide and chlorine and the liquid acrylonitrile, the WHAZAN dense cloud dispersion model was used. For the alternative-case release of allyl alcohol, the buoyant plume dispersion model was used for consistency with the guidance methodology. The release rates estimated using the guidance methodology were entered as the release rates for all of the WHAZAN modeling.

The WHAZAN dense cloud dispersion requires a "volume dilution factor" as one of its inputs. This factor was not explained; it was presumed to account for dilution of pressurized gases with air upon release. For the gases modeled, the default dilution factor of 60 was used; for acrylonitrile, a dilution factor of 0 was entered. This factor appears to have little effect on the distance results.

D.5 Worst-Case Consequence Analysis for Flammable Substances

The equation used for the vapor cloud explosion analysis for the worst case involving flammable substances is given in Appendix C. This equation is based on the TNT-equivalency method of the UK Health and Safety Executive, as presented in the publication of the Center for Chemical Process Safety of the American Institute of Chemical Engineers (AIChE), *Guidelines for Evaluating the Characteristics of Vapor*

Cloud Explosions, Flash Fires, and BLEVEs (1994). The assumption was made for the worst case that the total quantity of the released substance is in the flammable part of the cloud. The AIChE document lists this assumption as one of a number that have been used for vapor cloud explosion blast prediction; it was chosen as a conservative assumption for the worst-case analysis. The yield factor of 10 percent was a conservative worst-case assumption, based on information presented in the AIChE document. According to the AIChE document, reported values for TNT equivalency for vapor cloud explosions range from a fraction of one percent to tens of percent; for most major vapor cloud explosions, the range is one to ten percent.

The endpoint for the vapor cloud explosion analysis, 1 psi, is reported to cause damage such as shattering of glass windows and partial demolition of houses. Skin laceration from flying glass also is reported. This endpoint was chosen for the consequence analysis because of the potential for serious injuries to people from the property damage that might result from an explosion.

The TNT equivalent model was chosen as the basis for the consequence analysis because of its simplicity and wide use. This model does not take into account site-specific factors and many chemical-specific factors that may affect the results of a vapor cloud explosion. Other methods are available for vapor cloud explosion modeling; see the list of references in Appendix A for some publications that include information on other vapor cloud explosion modeling methods.

D.6 Alternative Scenario Analysis for Gases

The equation for estimating release rate of a gas from a hole in a tank is based on the equations for gas discharge rate presented in the *Handbook of Chemical Hazard Analysis Procedures* by the Federal Emergency Management Agency (FEMA), DOT, and EPA, and equations in EPA's *Workbook of Screening Techniques for Assessing Impacts of Toxic Air Pollutants*. The equation for an instantaneous discharge under non-choked flow conditions is:

$$m = C_d\, A_h \sqrt{2 p_0 \rho_0 \left(\frac{\gamma}{\gamma - 1}\right) \left[\left(\frac{p_1}{p_0}\right)^{\frac{2}{\gamma}} - \left(\frac{p_1}{p_0}\right)^{\frac{\gamma+1}{\gamma}}\right]} \qquad \text{(D-7)}$$

where:

m	=	Discharge rate (kg/s)
C_d	=	Discharge coefficient
A_h	=	Opening area (m^2)
γ	=	Ratio of specific heats
p_0	=	Tank pressure (Pascals)
p_1	=	Ambient pressure (Pascals)
ρ_0	=	Density (kg/m^3)

Under choked flow conditions (maximum flow rate), the equation becomes:

$$m = C_d \, A_h \sqrt{\gamma p_0 \rho_0 \left(\frac{2}{\gamma + 1}\right)^{\frac{\gamma+1}{\gamma-1}}} \qquad \textbf{(D-8)}$$

For development of the equation and gas factors presented in this guidance, density (ρ) was rewritten as a function of pressure and molecular weight, based on the ideal gas law:

$$\rho = \frac{p_0 \, MW}{R \; T_t} \qquad \textbf{(D-9)}$$

where:

MW	=	Molecular weight (kilograms per kilomole)
R	=	Gas constant (8,314 Joules per degree-kilomole)
T_t	=	Tank temperature (K)

The choked flow equation can be rewritten:

$$m = C_d \, A_h \, p_0 \, \frac{1}{\sqrt{T_t}} \sqrt{\gamma \left(\frac{2}{\gamma + 1}\right)^{\frac{\gamma+1}{\gamma-1}}} \sqrt{\frac{MW}{8314}} \qquad \textbf{(D-10)}$$

To derive the equation presented in the guidance, all the chemical-specific properties, constants, and appropriate conversion factors were combined into the "Gas Factor" (GF). The discharge coefficient was assumed to have a value of 0.8, based on the screening value recommended in EPA's *Workbook of Screening Techniques for Assessing Impacts of Toxic Air Pollutants*. The GF was derived as follows:

$$GF = 132.2 \times 6{,}895 \times 6.4516 \times 10^{-4} \times 0.8 \sqrt{\gamma \left(\frac{2}{\gamma + 1}\right)^{\frac{\gamma+1}{\gamma-1}}} \sqrt{\frac{MW}{8314}} \qquad \textbf{(D-11)}$$

where:

132.2	=	Conversion factor for lbs/min to kg/s
6,895	=	Conversion factor for psi to Pascals (p_0)
6.4516×10^{-4}	=	Conversion factor for square inches to square meters (A_h)

GF values were calculated for all gases regulated under CAA section 112(r) and are listed in Appendix B, Exhibit B-1, for toxic gases and Appendix C, Exhibit C-2, for flammable gases.

From the equation for choked flow above and the equation for the GF above, the initial release rate for a gas from a hole in a tank can be written as:

$$QR = HA \times P_t \times \frac{1}{\sqrt{T_t}} \times GF \qquad \textbf{(D-12)}$$

where:			
	QR	=	Release rate (pounds per minute)
	HA	=	Hole area (square inches)
	P_t	=	Tank pressure (psia)
	T_t	=	Tank temperature (K)

D.7 Alternative Scenario Analysis for Liquids

D.7.1 Releases from Holes in Tanks

The equation for estimating release rate of a liquid from a hole in a tank is based on the equations for liquid release rate presented in the *Handbook of Chemical Hazard Analysis Procedures* by FEMA, DOT, and EPA and EPA's *Workbook of Screening Techniques for Assessing Impacts of Toxic Air Pollutants*. The equation for the instantaneous release rate is:

$$m = A_h C_d \sqrt{\rho_l \left[2g\rho_l \left(H_L - H_h\right) + 2\left(P_0 - P_a\right)\right]} \qquad \textbf{(D-13)}$$

where:			
	m	=	Discharge rate (kilograms per second)
	A_h	=	Opening area (square meters)
	C_d	=	Discharge coefficient (unitless)
	g	=	Gravitational constant (9.8 meters per second squared)
	ρ_l	=	Liquid density (kilograms per cubic meter)
	P_0	=	Storage pressure (Pascals)
	P_a	=	Ambient pressure (Pascals)
	H_L	=	Liquid height above bottom of container (meters)
	H_h	=	Height of opening (meters)

A version of this equation is presented in the guidance for use with data found in Appendix B, for gases liquefied under pressure. The equation in the text was derived using the conversion factors listed below and density factors and equilibrium vapor pressure or tank pressure values listed in Appendix B, Exhibit B-1. Equation D-13 becomes:

$$QR = 132.2 \times 6.4516 \times 10^{-4} \times 0.8 \times HA \sqrt{16.018 \times d \times [2 \times 9.8 \times 16.018 \times d \times LH \times 0.0254 + 2P_g \times 6895]} \qquad \textbf{(D-14)}$$

where:			
where:	QR	=	Release rate (pounds per minute)
	HA	=	Hole area (square inches)
	132.2	=	Conversion factor for kilograms per second to pounds per minute
	6.4516×10^{-4}	=	Conversion factor for square inches to square meters (HA)
	0.8	=	Discharge coefficient (0.8)
	d	=	Liquid density (pounds per cubic foot); can derived by using the density factor: 1/(DFx0.033)
	16.018	=	Conversion factor for pounds per cubic feet to kilograms per cubic meters (D)
	9.8	=	Gravitational constant (meters per second squared)
	LH	=	Height of liquid above hole (inches)
	2.54×10^{-2}	=	Conversion factor for inches to meters (LH)
	P_g	=	Gauge pressure in tank (psi)
	6,895	=	Conversion factor for psi to Pascals (P_g)

After combining the conversion factors and incorporating the density factor (DF), this equation becomes:

$$QR = HA \times 6.82 \sqrt{\frac{0.7}{DF^2} \times LH + \frac{669}{DF} \times P_g} \qquad \textbf{(D-15)}$$

For liquids stored at ambient pressure, Equation D-13 becomes:

$$m = A_h C_d \rho_l \sqrt{2g\,(H_L - H_h)} \qquad \textbf{(D-16)}$$

To derive the equation presented in the guidance for liquids under ambient pressure, all the chemical-specific properties, constants, and conversion factors were combined into the "Liquid Leak Factor" (LLF). The discharge coefficient was assumed to have a value of 0.8, based on the screening value recommended in EPA's *Workbook of Screening Techniques for Assessing Impacts of Toxic Air Pollutants*. The LLF was derived as follows:

$$LLF = 132.2 \times 6.4516 \times 10^{-4} \times 0.1594 \times 0.8 \times \sqrt{2 \times 9.8} \times \rho_l \qquad \textbf{(D-17)}$$

where:			
where:	LLF	=	Liquid Leak Factor (pounds per minute-inches$^{2.5}$)
	132.2	=	Conversion factor for kilograms per second to pounds per minute (m)
	6.4516×10^{-4}	=	Conversion factor for square inches to square meters (A_h)
	0.1594	=	Conversion factor for square root of inches to square root of meters (H_L - H_h)
	0.8	=	Discharge coefficient (0.8)
	9.8	=	Gravitational constant (meters per second squared)
	ρ_l	=	Liquid density (kilograms per cubic meter)

LLF values were calculated for all liquids regulated under CAA section 112(r) and are listed in Appendix B, Exhibit B-2, for toxic liquids and Appendix C, Exhibit C-3, for flammable liquids.

From the equation for liquid release rate from a hole in a tank at ambient pressure and the equation for the LLF, the initial release rate for a liquid from a tank under atmospheric pressure can be written as:

$$QR_L = HA \times \sqrt{LH} \times LLF \qquad \text{(D-18)}$$

where:

QR_L	=	Liquid release rate (pounds per minute)
HA	=	Hole area (square inches)
LH	=	Height of liquid above hole (inches)

D.7.2 Releases from Pipes

The equation used to estimate releases of liquids from pipes is the Bernoulli equation. It assumes that the density of the liquid is constant and does not account for losses in velocity due to wall friction. The equation follows:

$$\frac{(P_a - P_b)}{D} + \frac{g\ (Z_a - Z_b)}{g_c} = \frac{(V_b^2 - V_a^2)}{2g_c} \qquad \text{(D-19)}$$

where:

P_a	=	Pressure at pipe inlet (Pascals)
P_b	=	Pressure at pipe outlet (Pascals)
Z_a	=	Height above datum plane at pipe inlet (meters)
Z_b	=	Height above datum plane at pipe release (meters)
g	=	Gravitational acceleration (9.8 meters per second squared)
g_c	=	Newton's law proportionality factor (1.0)
V_a	=	Operational velocity (meters per second)
V_b	=	Release velocity (meters per second)
D	=	Density of liquid (kilograms per cubic meter)

Isolating V_b yields:

$$V_b = \sqrt{\frac{2\ g_c\ (P_a - P_b)}{D} + 2\ g\ (Z_a - Z_b) + V_a^2} \qquad \text{(D-20)}$$

To develop the equation presented in the text, conversion factors for English units and constants were incorporated as follows:

$$V_b = 197\sqrt{\frac{2\times 6895\times(P_T - 14.7)\times DF\times 0.033}{16.08} + (2\times 9.8\times 0.3048\times(Z_a - Z_b) + 0.00508^2\times V_a^2} \quad \textbf{(D-21)}$$

where:

V_b	=	Release velocity (feet per minute)
197	=	Conversion factor for meters per second to feet per minute
6895	=	Conversion factor for psi to Pascals
P_T	=	Total pipe pressure (psi)
14.7	=	Atmospheric pressure (psi)
16.08	=	Conversion factor for pounds per cubic foot to kilograms per cubic meter
DF	=	Density factor (1/(0.033 DF)= density in pounds per cubic foot)
9.8	=	Gravitational acceleration (meters per second^2)
0.3048	=	Conversion factor for feet to meters
Z_a-Z_b	=	Change in pipe elevation, inlet to outlet (feet)
0.00508	=	Conversion factor for feet per minute to meters per second
V_a	=	Operational velocity (feet per minute)

D.8 Vapor Cloud Fires

Factors for leaks from tanks for flammable substances (GF and LLF) were derived as described for toxic substances (see above).

The endpoint for estimating impact distances for vapor cloud fires of flammable substances is the lower flammability limit (LFL). The LFL is one of the endpoints for releases of flammable substances specified in the RMP Rule. It was chosen to provide a reasonable, but not overly conservative, estimation of the possible extent of a vapor cloud fire.

D.9 Pool Fires

A factor used for estimating the distance to a heat radiation level from a pool fire that could cause second degree burns from a 40-second exposure was developed based on equations presented in the AIChE document, *Guidelines for Evaluating the Characteristics of Vapor Cloud Explosions, Flash Fires, and BLEVEs* and in the Netherlands TNO document, *Methods for the Determination of Possible Damage to People and Objects Resulting from Releases of Hazardous Materials* (1992). The AIChE and TNO documents present a point-source model that assumes that a selected fraction of the heat of combustion is emitted as radiation in all directions. The radiation per unit area received by a target at some distance from the point source is given by:

$$q = \frac{f\ m\ H_c\ \tau_a}{4\pi x^2} \quad \textbf{(D-22)}$$

where:

q	=	Radiation per unit area received by the receptor (Watts per square meter)
m	=	Rate of combustion (kilograms per second)
τ_a	=	Atmospheric transmissivity
H_c	=	Heat of combustion (Joules per kilogram)
f	=	Fraction of heat of combustion radiated
x	=	Distance from point source to receptor (meters)

The fraction of combustion energy dissipated as thermal radiation (f in the equation above) is reported to range from 0.1 to 0.4. To develop factors for estimating distances for pool fires, this fraction was assumed to be 0.4 for all the regulated flammable substances. The heat radiation level (q) was assumed to be 5 kilowatts (5,000 Watts) per square meter. This level is reported to cause second degree burns from a 40-second exposure. One of the endpoints for releases of flammable substances specified in the RMP Rule is 5 kilowatts per square meter for 40 seconds. It was assumed that people would be able to escape from the heat in 40 seconds. The atmospheric transmissivity (τ_a) was assumed equal to one.

For a pool fire of a flammable substance with a boiling point above the ambient temperature, the combustion rate can be estimated by the following empirical equation:

$$m = \frac{0.0010\ H_c\ A}{H_v + C_p\ (T_b - T_a)} \qquad \textbf{(D-23)}$$

where:

m	=	Rate of combustion (kilograms per second)
H_c	=	Heat of combustion (Joules per kilogram)
H_v	=	Heat of vaporization (Joules per kilogram)
C_p	=	Liquid heat capacity (Joules per kilogram-degree K)
A	=	Pool area (square meters)
T_b	=	Boiling temperature (K)
T_a	=	Ambient temperature (K)
0.0010	=	Constant

Combining Equations D-22 and D-23 (above), and assuming a heat radiation level of 5,000 Watts per square meter, gives the following equation for liquid pools of substances with boiling points above ambient temperature:

$$x = H_c \sqrt{\frac{0.4 \left(\dfrac{0.0010\ A}{H_v + C_p(T_b - T_a)}\right)}{4\pi q}} \qquad \textbf{(D-24)}$$

or

$$x = H_c \sqrt{\frac{0.0001\ A}{5{,}000\pi\ (H_v + C_p(T_b - T_a))}} \qquad \textbf{(D-25)}$$

where:	x	=	Distance from point source to receptor (meters)
	q	=	Radiation per unit area received by the receptor = 5,000 Watts per square meter
	H_c	=	Heat of combustion (Joules per kilogram)
	f	=	Fraction of heat of combustion radiated = 0.4
	H_v	=	Heat of vaporization (Joules per kilogram)
	C_p	=	Liquid heat capacity (Joules per kilogram-degree Kelvin)
	A	=	Pool area (square meters)
	T_b	=	Boiling temperature (K)
	T_a	=	Ambient temperature (K)
	0.0010	=	Constant

For a pool fire of a flammable substance with a boiling point below the ambient temperature (i.e., liquefied gases) the combustion rate can be estimated by the following equation, based on the TNO document:

$$m = \frac{0.0010\ H_c\ A}{H_v} \qquad \textbf{(D-26)}$$

where:	m	=	Rate of combustion (kilograms per second)
	H_v	=	Heat of vaporization (Joules per kilogram)
	H_c	=	Heat of combustion (Joules per kilogram)
	A	=	Pool area (square meters)
	0.0010	=	Constant

Then the equation for distance at which the radiation received equals 5,000 Watts per square meter becomes:

$$x = H_c \sqrt{\frac{0.0001\ A}{5{,}000\pi\ H_v}} \qquad \textbf{(D-27)}$$

where:	x	=	Distance from point source to receptor (meters)
	5,000	=	Radiation per unit area received by the receptor (Watts per square meter)
	H_c	=	Heat of combustion (Joules per kilogram)
	H_v	=	Heat of vaporization (Joules per kilogram)
	A	=	Pool area (square meters)
	0.0001	=	Derived constant (see equations D-20 and D-21)

A "Pool Fire Factor" (PFF) was calculated for each regulated flammable liquid and gas (to be applied to gases liquefied by refrigeration) to allow estimation of the distance to the heat radiation level that would lead to second degree burns. For the derivation of this factor, ambient temperature was assumed to be 298 K (25 °C). Other factors are discussed above. The PFF for liquids with boiling points above ambient temperature was derived as follows:

$$PFF = H_c \sqrt{\frac{0.0001}{5{,}000\pi\ [H_v + C_p(T_b - 298)]}} \qquad \textbf{(D-28)}$$

where:	5,000	=	Radiation per unit area received by the receptor (Watts per square meter)
	H_c	=	Heat of combustion (Joules per kilogram)
	H_v	=	Heat of vaporization (Joules per kilogram)
	C_p	=	Liquid heat capacity (Joules per kilogram-degree K)
	T_b	=	Boiling temperature (K)
	298	=	Assumed ambient temperature (K)
	0.0001	=	Derived constant (see above)

For liquids with boiling points below ambient temperature, the PFF is derived as follows:

$$PFF = H_c \sqrt{\frac{0.0001}{5{,}000\pi\ H_v}} \qquad \textbf{(D-29)}$$

where:	5,000	=	Radiation per unit area received by the receptor (Watts per square meter)
	H_c	=	Heat of combustion (Joules per kilogram)
	H_v	=	Heat of vaporization (Joules per kilogram)
	0.0001	=	Derived constant (see above)

Distances where exposed people could potentially suffer second degree burns can be estimated as the PFF multiplied by the square root of the pool area (in square feet), as discussed in the text.

D.10 BLEVEs

Reference Table 30, the table of distances for BLEVEs, was developed based on equations presented in the AIChE document, *Guidelines for Evaluating the Characteristics of Vapor Cloud Explosions, Flash Fires, and BLEVEs*. The Hymes point-source model for a fireball, as cited in the AIChE document, uses the following equation for the radiation received by a receptor:

$$q = \frac{2.2\ \tau_a\ R\ H_c\ m_f^{0.67}}{4\pi L^2} \qquad \textbf{(D-30)}$$

where:	q	=	Radiation received by the receptor (Watts per square meter)
	m_f	=	Mass of fuel in the fireball (kg)

τ_a	=	Atmospheric transmissivity
H_c	=	Heat of combustion (Joules per kilogram)
R	=	Radiative fraction of heat of combustion
L	=	Distance from fireball center to receptor (meters)
π	=	3.14

Hymes (as cited by AIChE) suggests the following values for R:

R	=	0.3 for vessels bursting below relief valve pressure
R	=	0.4 for vessels bursting at or above relief valve pressure

For development of Reference Table 30, the following conservative assumptions were made:

R	=	0.4
τ_a	=	1

The effects of radiant heat on an exposed person depend on both the intensity of the radiation and the duration of the exposure. For development of the table of distances for BLEVEs, it was assumed that the time of exposure would equal the duration of the fireball. The AIChE document gives the following equations for duration of a fireball:

$$t_c = 0.45\ m_f^{1/3}\ for\ m_f < 30{,}000\ kg \tag{D-31}$$

and

$$t_c = 2.6\ m_f^{1/6}\ for\ m_f > 30{,}000\ kg \tag{D-32}$$

where:

m_f	=	Mass of fuel (kg)
t_c	=	Combustion duration (seconds)

According to several sources (e.g., Eisenberg, et al., *Vulnerability Model, A Simulation System for Assessing Damage Resulting from Marine Spills*; Mudan, *Thermal Radiation Hazards from Hydrocarbon Pool Fires* (citing K. Buettner)), the effects of thermal radiation are generally proportional to radiation intensity to the four-thirds power times time of exposure. Thus, a thermal "dose" can be estimated using the following equation:

$$Dose = t\ q^{4/3} \tag{D-33}$$

where:

t	=	Duration of exposure (seconds)
q	=	Radiation intensity (Watts/m^2)

The thermal "dose" that could cause second-degree burns was estimated assuming 40 seconds as the duration of exposure and 5,000 Watts/m^2 as the radiation intensity. The corresponding dose is 3,420,000 (Watts/m^2)$^{4/3}$-second.

For estimating the distance from a fireball at which a receptor might receive enough thermal radiation to cause second degree burns, the dose estimated above was substituted into the equation for radiation received from a fireball:

$$q = \left[\frac{3{,}420{,}000}{t}\right]^{\frac{3}{4}} \quad \textbf{(D-34)}$$

$$\left[\frac{3{,}420{,}000}{t}\right]^{3/4} = \frac{2.2\ \tau_a\ R\ H_c\ m_f^{0.67}}{4\pi L^2} \quad \textbf{(D-35)}$$

$$L = \sqrt{\frac{2.2\ \tau_a\ R\ H_c\ m_f^{0.67}}{4\pi \left[\frac{3{,}420{,}000}{t}\right]^{3/4}}} \quad \textbf{(D-36)}$$

where:

where:	L	=	Distance from fireball center to receptor (meters)
	q	=	Radiation received by the receptor (Watts per square meter)
	m_f	=	Mass of fuel in the fireball (kg)
	τ_a	=	Atmospheric transmissivity (assumed to be 1)
	H_c	=	Heat of combustion (Joules per kilogram)
	R	=	Radiative fraction of heat of combustion (assumed to be 0.4)
	t	=	Duration of the fireball (seconds) (estimated from the equations above); assumed to be duration of exposure

Equation D-36 was used to develop the reference table for BLEVEs presented in the text (Reference Table 30).

D.11 Alternative Scenario Analysis for Vapor Cloud Explosions

According to T.A. Kletz, in "Unconfined Vapor Cloud Explosions" (Eleventh Loss Prevention Symposium, sponsored by AIChE, 1977), unconfined vapor cloud explosions almost always result from the release of flashing liquids. For this reason, the quantity in the cloud for the alternative scenario vapor cloud explosion in this guidance is based on the fraction flashed from the release of a flammable gas liquefied under pressure. The guidance provides a method to estimate the quantity in the cloud from the fraction flashed into vapor plus the quantity that might be carried along as aerosol. The recommendation to use twice the quantity flashed as the mass in the cloud (so long as it does not exceed the total amount of flammable substance available) is based on the method recommended by the UK Health and Safety Executive (HSE), as cited in the AIChE document, *Guidelines for Evaluating the Characteristics of Vapor Cloud Explosions, Flash Fires, and BLEVEs*. The factor of two is intended to allow for spray and aerosol formation.

The equation for the flash fraction, for possible use in for the alternative scenario analysis, is based on the Netherlands TNO document, *Methods for the Calculation of the Physical Effects of the Escape of Dangerous Material* (1980), Chapter 4, "Spray Release." The following equation is provided:

$$X_{vap,a} = \left(X_{vap,b}\ \frac{T_b}{T_l}\right) + \left(\frac{T_b C_l}{h_v}\ \ln\frac{T_l}{T_b}\right) \qquad \textbf{(D-37)}$$

where: $X_{vap,a}$ = Weight fraction of vapor after expansion
$X_{vap,b}$ = Weight fraction of vapor before expansion (assumed to be 0 for calculation of the flash fraction)
T_b = Boiling temperature of gas compressed to liquid (K)
T_l = Temperature of stored gas compressed to liquid (K)
C_l = Specific heat of gas compressed to liquid (Joules/kilogram-K)
h_v = Heat of evaporation of gas compressed to liquid (Joules/kilogram)

To develop a Flash Fraction Factor (FFF) for use in consequence analysis, compressed gases were assumed to be stored at 25 °C (298 K) (except in cases where the gas could not be liquefied at that temperature). The equation for FFF is:

$$FFF = \left(\frac{T_b C_l}{h_v}\ \ln\frac{298}{T_b}\right) \qquad \textbf{(D-38)}$$

where: T_b = Boiling temperature of gas compressed to liquid (K)
C_l = Specific heat of gas compressed to liquid (Joules/kilogram-K)
h_v = Heat of evaporation of gas compressed to liquid (Joules/kilogram)
298 = Temperature of stored gas compressed to liquid (K)

The recommendation to use a yield factor of 0.03 for the alternative scenario analysis for vapor cloud explosions also is based on the UK HSE method cited by AIChE. According to the AIChE document, this recommendation is based on surveys showing than most major vapor cloud explosions have developed between 1 percent and 3 percent of available energy.

APPENDIX E

WORKSHEETS FOR OFFSITE CONSEQUENCE ANALYSIS

Using the Methods in this Guidance

WORKSHEET 1
WORST-CASE ANALYSIS FOR TOXIC GAS

1. Select Scenario (*defined by rule for worst case as release of largest quantity over 10 minutes*)		**Guidance Reference**
• Identify toxic gas	*Name:* ____________________ *CAS number:* ________-____-___	Chapter 2 Section 3.1
• Identify largest quantity in largest vessel or pipeline	*Quantity (pounds):* ________	
• Identify worst-case meteorological conditions	*Atmospheric stability class:* F *Wind speed:* 1.5 m/s *Ambient temperature:* 25 °C *Relative humidity:* 50%	
2. Determine Release Rate		
• Estimate release rate *Quantity/10 min, except gases liquefied by refrigeration in some cases*	*Release rate (lbs/min):* __________ *Will release always take place in enclosure?* ___ (If yes, go to next step)	Section 3.1.1
• Revise release rate to account for passive mitigation (enclosure)	*Can release cause failure of enclosure?* ___ (If yes, use unmitigated release rate) *Factor to account for enclosure:* 0.55 *Mitigated release rate (lbs/min):* ________	Section 3.1.2
3. Determine Distance to the Endpoint Specified by Rule		
• Identify endpoint	*Endpoint (mg/L):* __________	Exhibit B-1
• Determine gas density *Consider conditions (e.g., liquefied under pressure)*	*Dense:* ___ *Neutrally buoyant:* ___	Exhibit B-1
• Determine site topography *Rural and urban defined by rule*	*Rural:* ___ *Urban:* ___	Section 2.1
• Determine appropriate reference table of distances *Use 10-minute tables*	*Reference table used (number):* ___	Chapter 4 Reference Tables 1-12
• Find distance on reference table	*Release rate/endpoint (neutrally buoyant):* ____ *Distance to endpoint (mi):* ______	Chapter 4 Reference Tables 1-12

WORKSHEET 2
WORST-CASE ANALYSIS FOR TOXIC LIQUID

1. Select Scenario *(defined by rule for worst case as release of largest quantity to form an evaporating pool)*		**Guidance Reference**
• Identify toxic liquid • Identify concentration for solutions or mixtures	*Name:* ____________________ *CAS number:* ________-_____-___ *Concentration in solution or mixture (wt %):* ___	Chapter 2 Section 3.2 Section 3.2.4 for mixtures
• Identify largest quantity in largest vessel or pipeline	*Quantity (pounds):* ________ *Quantity of regulated substance in mixture:* ____	
• Identify worst-case meteorological conditions	*Atmospheric stability class:* F *Wind speed:* 1.5 m/s *Ambient temperature:* 25 °C *Relative humidity:* 50%	
2. Determine Release Rate		
• Determine temperature of spilled liquid *Must be highest maximum daily temperature or process temperature, or boiling point for gases liquefied by refrigeration*	*Temperature of liquid (°C):* _____	Section 3.2 Section 3.1.3
• Determine appropriate liquid factors for release rate estimation	*LFA:* _______ *LFB:* _______ *DF:* _______ *TCF:* _______	Section 3.2, Exhibits B-2, B-4 Section 3.3, Exhibit B-3 for water solutions
Estimate Maximum Pool Area		
• Estimate maximum pool area *Spilled liquid forms pool 1 cm deep*	*Maximum pool area (ft^2):* ______	Section 3.2.3 Equation 3-6

WORKSHEET 2 (continued)

Estimate Pool Area for Spill into Diked Area		
• Estimate diked area *Consider failure of dikes or overflow of diked area*	***Diked area*** *(ft^2):* _______ ***Is diked area smaller than maximum area?*** ____ (If no, use maximum area to estimate release rate) ***Diked volume*** *(ft^3):* _______ ***Spilled volume*** *(ft^3):* _______ ***Is spilled volume smaller than diked volume?***___ (If no, estimate overflow) ***Overflow volume*** *(ft^3):* _______ ***Overflow area*** *(ft^2):* _______	Section 3.2.3
• Choose pool area for release rate estimation *Maximum area, diked area, or sum of diked area and overflow area*	*Pool area (ft^2):* _______	Section 3.2.3
Estimate Release Rate from Pool		
• Estimate release rate for undiked pool (maximum pool area) *Based on quantity spilled, LFA or LFB, and DF*	***Release rate*** *(lbs/min):* _______	Section 3.2.2 Section 3.2.4 (mixtures) Equation 3-3 or 3-4
• Estimate release rate for diked pool (use pool area from previous section) *Based on pool area and LFA or LFB*	***Release rate*** *(lbs/min):* _______	Section 3.2.2 Section 3.2.4 (mixtures) Equation 3-7 or 3-8
• Revise release rate for release in building *Apply factor to release rate*	***Release rate if outside*** *(lbs/min)* _______ (Use release rate for undiked or diked pool) ***Factor to account for enclosure:*** 0.1 ***Revised release rate*** *(lbs/min):* _______	Section 3.2.3 Equations 3-9, 3-10
• Revise release rate for temperature *Apply appropriate TCF to release rate*	***Revised release rate*** *(lbs/min):* _______	Section 3.2.5 Equation 3-11
• Estimate duration of release	***Release duration*** *(min):* _______	Section 3.2.2 Equation 3-5

WORKSHEET 2 (continued)

3. Determine Distance to the Endpoint		
• Identify endpoint *Specified by rule*	*Endpoint (mg/L):* _______	Exhibit B-2
• Determine vapor density	*Dense:* ____ *Neutrally buoyant:* ____	Exhibit B-2
• Determine site topography *Rural and urban defined by rule*	*Rural:* ____ *Urban:* ____	Section 2.1
• Determine appropriate reference table of distances *Based on release duration, vapor density, topography*	*Reference table used (number):* ____	Chapter 4 Reference Tables 1-12
• Find distance on reference table	*Release rate/endpoint (neutrally buoyant):* _____ *Distance to endpoint (mi):* _____	Chapter 4 Reference Tables 1-12

WORKSHEET 3
WORST-CASE ANALYSIS FOR FLAMMABLE SUBSTANCE

1. Select Scenario *(defined by rule for worst case as vapor cloud explosion of largest quantity)*		**Guidance Reference**
• Identify flammable substance	*Name:* ____________ *CAS number:* ______-______-____	Chapter 2 Section 3.1
• Identify largest quantity in largest vessel or pipeline *Consider total quantity of flammable substance, including non-regulated substances in flammable mixtures*	*Quantity (pounds):* ______	
2. Determine Distance to the Endpoint *(endpoint specified by the rule as 1 psi overpressure; yield factor assumed to be 10% for TNT-equivalent model)*		
• Estimate distance to 1 psi using Reference Table *Find quantity, read distance from table*	*Distance to 1 psi (mi):* ______	Chapter 5 Reference Table 13
• Alternatively, estimate distance to 1 psi using equation	For pure substance: *Heat of combustion (kJ/kg):* ______ For mixture: *Heat of combustion of major component (kJ/kg):* ______ *Heats of combustion of other components (kJ/kg):* ______, ______, ______ *Distance to 1 psi (mi).* ______	Chapter 5 Appendix C.1 Appendix C.2 Exhibit C-1

WORKSHEET 4
ALTERNATIVE SCENARIO ANALYSIS FOR TOXIC GAS

1. Select Scenario		**Guidance Reference**
• Identify toxic gas	*Name:* __________ *CAS number:* ____-____-____	Chapter 6 Chapter 7 Section 7.1
• Identify conditions of storage or processing of toxic gas *Treat gases liquefied by refrigeration as liquids*	*Non-liquefied pressurized gas:* ____ *Gas liquefied under pressure:* ____ *In tank:* ____ *In pipeline:* ____ *Other (describe):* __________	
• Develop alternative scenario ▹ More likely than worst case ▹ Should reach endpoint off site	*Describe scenario:* __________	
• Identify average meteorological conditions	*Atmospheric stability class:* D *Wind speed:* 3.0 m/s *Ambient temperature:* 25 °C *Relative humidity:* 50%	
2. Determine Release Rate		
• Estimate gas release rate from hole in tank (choked/ maximum flow) for ▹ Pressurized gas ▹ Gas liquefied under pressure released from vapor space	*Hole area (in^2):* ____ *Tank pressure (psia):* ____ *Tank temperature (K):* ____ *GF:* ____ *Release rate (lbs/min):* ____	Section 7.1.1 Equation 7-1 Exhibit B-1
• Estimate flashing liquid release rate from hole in tank ▹ Gas liquefied under pressure released from liquid space	*Hole area (in^2):* ____ *Tank pressure (psig):* ____ *DF:* ____ *Liquid height above hole (in):* ____ *Release rate (lbs/min):* ____	Section 7.1.2 Equation 7-2 Exhibit B-1

WORKSHEET 4 (continued)

• Estimate flashing liquid release rate from break in long pipeline ▹ Gas liquefied under pressure completely filling pipeline	*Initial flow rate (lbs/min):* _______ *DF:* _______ *Initial flow velocity (ft/min):* _______ *Pipe pressure (psi):* _______ *Change in pipe elevation (ft):* _______ *Cross-sectional pipe area (ft^2):* _______ *Release rate (lbs/min):* _______	Sections 7.1.1 and 7.2.1 Exhibit B-1
• Estimate release duration	*Time to stop release (min):* _______ *Time to empty tank or pipe (min):* _______ *Default release duration:* 60 min	Section 7.1.1
• Revise release rate for passive mitigation (enclosure)	*Release rate if outside (lbs/min):* _______ *Factor to account for enclosure:* 0.55 *Revised release rate (lbs/min):* _______	Section 7.1.2 Section 3.1.2
• Revise release rate for active mitigation	*Active mitigation technique used:* ____________ ____________________ *Time to stop release using active technique (min):* _______ *Fractional release rate reduction by active technique:* _______ *Revised release rate (lb/min):* _______	
• Estimate release duration (mitigated release)	*Release duration (min):* _______	Section 7.1.2
• Other release rate estimation	*Release rate (lb/min):* _______ *Method of release rate estimation (describe):* ___ ____________________________ *Release duration (min):* _______	
3. Determine Distance to the Endpoint		
• Identify endpoint *Specified by rule*	*Endpoint (mg/L):* _______	Exhibit B-1
• Determine gas density *Consider conditions (e.g., liquefied under pressure, refrigeration)*	*Dense:* _____ *Neutrally buoyant:* _____	Exhibit B-1
• Determine site topography *Rural and urban defined by rule*	*Rural:* _____ *Urban:* _____	Section 2.1

WORKSHEET 4 (continued)

• Determine appropriate reference table of distances *Based on release duration, vapor density, and topography*	***Reference table used** (number):* _______	Chapter 8 Reference Tables 14-25
• Find distance on reference table	***Release rate/endpoint** (neutrally buoyant):* _____ ***Distance to endpoint** (mi):* _______	Chapter 8 Reference Tables 14-25

WORKSHEET 5
ALTERNATIVE SCENARIO ANALYSIS FOR TOXIC LIQUID

1. Select Scenario		**Guidance Reference**
• Identify toxic liquid *Include gases liquefied by refrigeration* • Identify concentration for solutions or mixtures	*Name:* ____________ *CAS number:* ______-_____-_____ *Concentration in solution or mixture (wt %):* ___	Chapter 6 Chapter 7 Section 7.2
• Identify conditions of storage or processing of toxic liquid	*Atmospheric tank:* ______ *Pressurized tank:* ______ *Pipeline:* ______ *Other (describe):* ____________ ____________	
• Develop alternative scenario ▹ More likely than worst case ▹ Should reach endpoint off site	*Describe scenario:* ____________ ____________ ____________	
• Identify meteorological conditions	*Atmospheric stability class:* F *Wind speed:* 3.0 m/s *Ambient temperature:* 25 °C *Relative humidity:* 50%	
2. Determine Release Rate		
Determine Liquid Release Rate and Quantity Released into Pool		
• Estimate liquid release rate from hole in atmospheric tank	*Hole area (in^2):* ______ *LLF:* ______ *Liquid height above hole (in):* ______ *Liquid release rate (lbs/min):* ______	Section 7.2.1 Equation 7-4 Exhibit B-2
• Estimate liquid release rate from break in long pipeline	*Initial flow rate (lbs/min):* ______ *DF:* ______ *Initial flow velocity (ft/min):* ______ *Pipe pressure (psi):* ______ *Change in pipe elevation (ft):* ______ *Cross-sectional pipe area (ft^2):* ______ *Liquid release rate (lbs/min):* ______	Section 7.2.1 Equations 7-5 - 7-7 Exhibit B-2
• Estimate liquid release duration	*Time to stop release (min):* ______ *Time to empty tank to level of hole (min):* _____	Section 7.2.1

WORKSHEET 5 (continued)

• Revise liquid release duration for active mitigation	*Active mitigation technique (describe):* ________ ________________________________ *Time to stop release (min):* _______	Section 7.2.2
• Estimate quantity of liquid released into pool *Liquid release rate times duration*	*Quantity of liquid released (lbs):* _______	Sections 7.2.1, 7.2.2, 7.2.3
Determine Pool Area and Evaporation Rate from Pool		
• Determine temperature of spilled liquid	*Temperature of liquid (°C):* _______	Section 7.2.3
• Determine appropriate liquid factors for release rate estimation	*LFA:* _______ *LFB:* _______ *DF:* _______ *TCF:* _______	Sections 7.2.3, 3.2, and Exhibits B-2, B-4 Section 3.3 and Exhibit B-3 for water solutions
Estimate Maximum Pool Area		
• Estimate maximum pool area *Spilled liquid forms pool 1 cm deep*	*Maximum pool area (ft²):* _______	Section 7.2.3, 3.2.3 Equation 3-6
Estimate Pool Area for Spill into Diked Area		
• Estimate diked area *Consider possibility of failure of dikes or overflow of diked area*	*Diked area (ft²):* _______ *Is diked area smaller than maximum area?* ____ (If no, use maximum area to estimate release rate) *Diked volume (ft³):* _______ *Spilled volume (ft³):* _______ *Is spilled volume smaller than diked volume?*___ (If no, estimate overflow) *Overflow volume (ft³):* _______ *Overflow area (ft²):* _______	Section 7.2.3, 3.2.3
• Choose pool area for evaporation rate estimation *Maximum area, diked area, or sum of diked area and overflow area*	*Pool area (ft²):* _______	Section 7.2.3, 3.2.3

WORKSHEET 5 (continued)

Estimate Release Rate from Pool		
• Estimate release rate for undiked pool *Based on quantity spilled, LFA or LFB, and DF*	*Release rate (lbs/min):* _______	Section 7.2 3 Section 3.2.4 (mixtures) Equation 7-8 or 7-9
• Estimate release rate for diked pool (use pool area from previous section) *Based on pool area and LFA or LFB*	*Release rate (lbs/min):* _______	Sections 7.2.3, 3.2.2 Section 3.2.4 (mixtures) Equation 7-10 or 7-11
• Revise release rate for temperature *Apply appropriate TCF to release rate*	*Revised release rate (lbs/min):* _______	Sections 7.2.3, 3.2.5 Equation 3-11
• Revise release rate for release in building *Apply factor to release rate*	*Release rate if outside (lbs/min):* _______ *Factor to account for enclosure:* 0.05 *Revised release rate (lbs/min):* _______	Sections 7.2.3, 3.2.3
• Revise release rate for active mitigation technique	*Active mitigation technique used:* ____________ ________________________________ *Fractional release rate reduction by active technique:* _______ *Revised release rate (lb/min):* _______	Section 7.2.3
• Compare liquid release rate and pool evaporation rate • Choose smaller release rate as release rate for analysis	*Release rate (lb/min):* _______	Section 7.2.3
3. Determine Distance to the Endpoint		
• Identify endpoint *Specified by rule*	*Endpoint (mg/L):* _______	Exhibit B-2
• Determine vapor density	*Dense:* _____ *Neutrally buoyant:* _____	Exhibit B-2
• Determine site topography *Rural and urban defined by rule*	*Rural:* _____ *Urban:* _____	Section 2.1

WORKSHEET 5 (continued)

• Determine appropriate reference table of distances *Based on release duration, vapor density, and topography*	***Reference table used*** *(number):* _______	Chapter 8 Reference Tables 14-25
• Find distance on reference table	***Release rate/endpoint*** *(neutrally buoyant):* _____ ***Distance to endpoint*** *(mi):* _______	Chapter 8 Reference Tables 14-25

WORKSHEET 6
ALTERNATIVE SCENARIO ANALYSIS FOR FLAMMABLE SUBSTANCE

1. Select Scenario		**Guidance Reference**
• Identify flammable substance	*Name:* ______________ *CAS number:* ______-_____-___	Chapter 6
• Identify conditions of storage or processing of flammable substance *Treat gases liquefied by refrigeration as liquids*	*Non-liquefied pressurized gas:* ______ *Gas liquefied under pressure:* ______ *Gas liquefied by refrigeration:* ______ *Liquid under atmospheric pressure:* ______ *Liquid under pressure greater than atmospheric:*_ *Other (describe):* ______________ ______________	
• Identify appropriate scenario ▹ Vapor cloud fire ▹ Pool fire ▹ BLEVE/fireball ▹ Vapor cloud explosion ▹ Other (not covered by OCA Guidance)	*Alternative scenario/type of fire or explosion (describe):* ______________ ______________ ______________	
2. Determine Release Rate		
Determine Release Rate for Vapor Cloud Fire		
• For gas releases and flashing liquid releases, *see Worksheet 4*	*Release rate (lbs/min):* ______	Section 9.1 Section 7.1 Equations 7-1, 7-2, 7-3 Exhibit C-2
• For liquid releases (non-flashing), *see Worksheet 5*	*Liquid release rate (lbs/min):* ______ *Liquid release duration (min):* ______ *Quantity in pool (lbs):* ______ *Release rate to air (lbs/min):* ______	Section 9.2 Section 7.2 Equations 7-4-7-12 Exhibit C-3
Determine Pool Area for Pool Fire		
Estimate pool area: *See Worksheet 5*	*Quantity in pool (lbs):* ______ *Pool area (ft^2):* ______	Sections 10.2 Section 7.2 Exhibits C-2, C-3

WORKSHEET 6 (continued)

Determine Quantity for BLEVE		
Determine quantity in tank	*Quantity (lbs):* _______	Section 10.3
Determine Quantity for Vapor Cloud Explosion		
Determine quantity in tank	*Quantity (lbs):* _______	Section 10.4
3. Determine Distance to the Endpoint		
• Identify endpoint suitable for scenario ▹ LFL ▹ 5 kW/m² for 40 seconds ▹ 1 psi overpressure	*Endpoint:* ______	Chapter 6 Exhibits C-2, C-3
Determine Distance to LFL for Vapor Cloud Fire		
• Determine vapor density	*Dense:* ____ *Neutrally buoyant:* _____	Exhibit B-2
• Determine site topography *Rural and urban defined by rule*	*Rural:* ____ *Urban:* ____	Section 2.1
• Determine appropriate reference table of distances *Based on vapor density and topography*	*Reference table used (number):* ______	Section 10.1 Reference Tables 26-29
• Find distance on reference table	*Release rate/endpoint (neutrally buoyant):* _____ *Distance to LFL (mi):* _______	Section 10.1 Reference Tables 26-29
Determine Distance to Heat Radiation Endpoint for Pool Fire		
• Calculate distance to 5 kW/m²	*PFF:* ______ *Pool area (ft²):* _______ *Distance (ft):* _______	Section 10.2 Equation 10-1

WORKSHEET 6 (continued)

Determine Distance to Heat Radiation Endpoint for BLEVE		
Determine distance for radiation from fireball equivalent to 5 kW/m² for 40 seconds	*Quantity (lbs):* _______ *Distance (mi):* _______	Section 10.3 Reference Table 30
Determine Distance to Overpressure Endpoint For Vapor Cloud Explosion		
Determine distance to 1 psi *Quantity in cloud can be less than total quantity* *Yield factor can be less than 10%*	*FFF:* _______ *Quantity flashed:* _______ *Yield factor:* _______ *Distance to 1 psi (mi):* _______	Section 10.4 Exhibit C-2 Reference Table 13

APPENDIX F

CHEMICAL ACCIDENT PREVENTION PROVISIONS

As codified at 40 CFR part 68 as of July 1, 1998

local agent, any noncompliance penalties owed by the source owner or operator shall be paid to the State or local agent.

APPENDIX A TO PART 67—TECHNICAL SUPPORT DOCUMENT

NOTE: EPA will make copies of appendix A available from: Director, Stationary Source Compliance Division, EN-341, 401 M Street, SW., Washington, DC 20460.

[54 FR 25259, June 20, 1989]

APPENDIX B TO PART 67—INSTRUCTION MANUAL

NOTE: EPA will make copies of appendix B available from: Director, Stationary Source Compliance Division, EN-341, 401 M Street, SW., Washington, DC 20460.

[54 FR 25259, June 20, 1989]

APPENDIX C TO PART 67—COMPUTER PROGRAM

NOTE: EPA will make copies of appendix C available from: Director, Stationary Source Compliance Division, EN-341, 401 M Street, SW., Washington, DC 20460.

[54 FR 25259, June 20, 1989]

PART 68—CHEMICAL ACCIDENT PREVENTION PROVISIONS

Subpart A—General

Subpart B—Hazard Assessment

Subpart C—Program 2 Prevention Program

Subpart D—Program 3 Prevention Program

Subpart E—Emergency Response

Subpart F—Regulated Substances for Accidental Release Prevention

Subpart G—Risk Management Plan

Subpart H—Other Requirements

AUTHORITY: 42 U.S.C. 7412(r), 7601(a)(1), 7661–7661f.

SOURCE: 59 FR 4493, Jan. 31, 1994, unless otherwise noted.

Subpart A—General

§ 68.1 Scope.

This part sets forth the list of regulated substances and thresholds, the petition process for adding or deleting substances to the list of regulated substances, the requirements for owners or operators of stationary sources concerning the prevention of accidental releases, and the State accidental release prevention programs approved under section 112(r). The list of substances, threshold quantities, and accident prevention regulations promulgated under this part do not limit in any way the general duty provisions under section 112(r)(1).

§ 68.2 Stayed provisions.

(a) Notwithstanding any other provision of this part, the effectiveness of the following provisions is stayed from March 2, 1994 to December 22, 1997.

(1) In Sec. 68.3, the definition of "stationary source," to the extent that such definition includes naturally occurring hydrocarbon reservoirs or transportation subject to oversight or regulation under a state natural gas or hazardous liquid program for which the state has in effect a certification to DOT under 49 U.S.C. 60105;

(2) Section 68.115(b)(2) of this part, to the extent that such provision requires an owner or operator to treat as a regulated flammable substance:

(i) Gasoline, when in distribution or related storage for use as fuel for internal combustion engines;

(ii) Naturally occurring hydrocarbon mixtures prior to entry into a petroleum refining process unit or a natural gas processing plant. Naturally occurring hydrocarbon mixtures include any of the following: condensate, crude oil, field gas, and produced water, each as defined in paragraph (b) of this section;

(iii) Other mixtures that contain a regulated flammable substance and that do not have a National Fire Protection Association flammability hazard rating of 4, the definition of which is in the NFPA 704, Standard System for the Identification of the Fire Hazards of Materials, National Fire Protection Association, Quincy, MA, 1990, available from the National Fire Protection Association, 1 Batterymarch Park, Quincy, MA 02269-9101; and

(3) Section 68.130(a).

(b) From March 2, 1994 to December 22, 1997, the following definitions shall apply to the stayed provisions described in paragraph (a) of this section:

Condensate means hydrocarbon liquid separated from natural gas that condenses because of changes in temperature, pressure, or both, and remains liquid at standard conditions.

Crude oil means any naturally occurring, unrefined petroleum liquid.

Field gas means gas extracted from a production well before the gas enters a natural gas processing plant.

Natural gas processing plant means any processing site engaged in the extraction of natural gas liquids from field gas, fractionation of natural gas liquids to natural gas products, or both. A separator, dehydration unit, heater treater, sweetening unit, compressor, or similar equipment shall not be considered a "processing site" unless such equipment is physically located within a natural gas processing plant (gas plant) site.

Petroleum refining process unit means a process unit used in an establishment primarily engaged in petroleum refining as defined in the Standard Industrial Classification code for petroleum refining (2911) and used for the following: Producing transportation fuels (such as gasoline, diesel fuels, and jet fuels), heating fuels (such as kerosene, fuel gas distillate, and fuel oils), or lubricants; separating petroleum; or separating, cracking, reacting, or reforming intermediate petroleum streams. Examples of such units include, but are not limited to, petroleum based solvent units, alkylation units, catalytic hydrotreating, catalytic hydrorefining, catalytic hydrocracking, catalytic reforming, catalytic cracking, crude distillation, lube oil processing, hydrogen production, isomerization, polymerization, thermal processes, and blending, sweetening, and treating processes. Petroleum refining process units include sulfur plants.

Produced water means water extracted from the earth from an oil or natural gas production well, or that is separated from oil or natural gas after extraction.

[59 FR 4493, Jan. 31, 1994, as amended at 61 FR 31731, June 20, 1996]

§ 68.3 Definitions.

For the purposes of this part:

Accidental release means an unanticipated emission of a regulated substance or other extremely hazardous substance into the ambient air from a stationary source.

Act means the Clean Air Act as amended (42 U.S.C. 7401 *et seq.*)

Administrative controls mean written procedural mechanisms used for hazard control.

Administrator means the administrator of the U.S. Environmental Protection Agency.

AIChE/CCPS means the American Institute of Chemical Engineers/Center for Chemical Process Safety.

API means the American Petroleum Institute.

Article means a manufactured item, as defined under 29 CFR 1910.1200(b), that is formed to a specific shape or design during manufacture, that has end use functions dependent in whole or in part upon the shape or design during end use, and that does not release or otherwise result in exposure to a regulated substance under normal conditions of processing and use.

ASME means the American Society of Mechanical Engineers.

CAS means the Chemical Abstracts Service.

Catastrophic release means a major uncontrolled emission, fire, or explosion, involving one or more regulated substances that presents imminent and substantial endangerment to public health and the environment.

Classified information means "classified information" as defined in the Classified Information Procedures Act, 18 U.S.C. App. 3, section 1(a) as "any information or material that has been determined by the United States Government pursuant to an executive order, statute, or regulation, to require protection against unauthorized disclosure for reasons of national security."

Condensate means hydrocarbon liquid separated from natural gas that condenses due to changes in temperature, pressure, or both, and remains liquid at standard conditions.

Covered process means a process that has a regulated substance present in more than a threshold quantity as determined under §68.115.

Crude oil means any naturally occurring, unrefined petroleum liquid.

Designated agency means the state, local, or Federal agency designated by the state under the provisions of §68.215(d) .

DOT means the United States Department of Transportation.

Environmental receptor means natural areas such as national or state parks, forests, or monuments; officially designated wildlife sanctuaries, preserves, refuges, or areas; and Federal wilderness areas, that could be exposed at any time to toxic concentrations, radiant heat, or overpressure greater than or equal to the endpoints provided in §68.22(a) , as a result of an accidental release and that can be identified on local U. S. Geological Survey maps.

Field gas means gas extracted from a production well before the gas enters a natural gas processing plant.

Hot work means work involving electric or gas welding, cutting, brazing, or similar flame or spark-producing operations.

Implementing agency means the state or local agency that obtains delegation for an accidental release prevention program under subpart E, 40 CFR part 63. The implementing agency may, but is not required to, be the state or local air permitting agency. If no state or local agency is granted delegation, EPA will be the implementing agency for that state.

Injury means any effect on a human that results either from direct exposure to toxic concentrations; radiant heat; or overpressures from accidental releases or from the direct consequences of a vapor cloud explosion (such as flying glass, debris, and other projectiles) from an accidental release and that requires medical treatment or hospitalization.

Major change means introduction of a new process, process equipment, or regulated substance, an alteration of process chemistry that results in any change to safe operating limits, or other alteration that introduces a new hazard.

Mechanical integrity means the process of ensuring that process equipment is fabricated from the proper materials of construction and is properly installed, maintained, and replaced to prevent failures and accidental releases.

Medical treatment means treatment, other than first aid, administered by a physician or registered professional personnel under standing orders from a physician.

Mitigation or mitigation system means specific activities, technologies, or equipment designed or deployed to capture or control substances upon loss of containment to minimize exposure of the public or the environment. Passive mitigation means equipment, devices, or technologies that function without human, mechanical, or other energy input. Active mitigation means equipment, devices, or technologies that need human, mechanical, or other energy input to function.

NFPA means the National Fire Protection Association.

Natural gas processing plant (gas plant) means any processing site engaged in the extraction of natural gas liquids from field gas, fractionation of mixed natural gas liquids to natural gas products, or both, classified as North American Industrial Classification System (NAICS) code 211112 (previously Standard Industrial Classification (SIC) code 1321).

Offsite means areas beyond the property boundary of the stationary source, and areas within the property boundary to which the public has routine and unrestricted access during or outside business hours.

OSHA means the U.S. Occupational Safety and Health Administration. Owner or operator means any person who owns, leases, operates, controls, or supervises a stationary source.

Petroleum refining process unit means a process unit used in an establishment primarily engaged in petroleum refining as defined in NAICS code 32411 for petroleum refining (formerly SIC code 2911) and used for the following: Producing transportation fuels (such as gasoline, diesel fuels, and jet fuels), heating fuels (such as kerosene, fuel gas distillate, and fuel oils), or lubricants; Separating petroleum; or Separating, cracking, reacting, or reforming intermediate petroleum streams. Examples of such units include, but are not limited to, petroleum based solvent units, alkylation units, catalytic hydrotreating, catalytic hydrorefining, catalytic hydrocracking, catalytic reforming, catalytic cracking, crude distillation, lube oil processing, hydrogen production, isomerization, polymerization, thermal processes, and blending, sweetening, and treating processes. Petroleum refining process units include sulfur plants.

Population means the public.

Process means any activity involving a regulated substance including any use, storage, manufacturing, handling, or on-site movement of such substances, or combination of these activities. For the purposes of this definition, any group of vessels that are interconnected, or separate vessels that are located such that a regulated substance could be involved in a potential release, shall be considered a single process.

Produced water means water extracted from the earth from an oil or natural gas production well, or that is separated from oil or natural gas after extraction.

Public means any person except employees or contractors at the stationary source.

Public receptor means offsite residences, institutions (e.g., schools, hospitals), industrial, commercial, and office buildings, parks, or recreational areas inhabited or occupied by the public at any time without restriction by the stationary source where members of the public could be exposed to toxic concentrations, radiant heat, or overpressure, as a result of an accidental release.

Regulated substance is any substance listed pursuant to section 112(r)(3) of the Clean Air Act as amended, in §68.130.

Replacement in kind means a replacement that satisfies the design specifications.

RMP means the risk management plan required under subpart G of this part.

SIC means Standard Industrial Classification.

Stationary source means any buildings, structures, equipment, installations, or substance emitting stationary activities which belong to the same industrial group, which are located on one or more contiguous properties, which are under the control of the same person (or persons under common control), and from which an accidental release may occur. The term stationary source does not apply to transportation, including storage incident to transportation, of any regulated substance or any other extremely hazardous substance under the provisions of this part. A stationary source includes transportation containers used for storage not incident to transportation and transportation containers connected to equipment at a stationary source for loading or unloading. Transportation includes, but is not limited to, transportation subject to oversight or regulation under 49 CFR parts 192, 193, or 195, or a state natural gas or hazardous liquid program for which the state has in effect a certification to DOT under 49 U.S.C. section 60105. A stationary source does not include naturally occurring hydrocarbon reservoirs. Properties shall not be considered contiguous solely because of a railroad or pipeline right-of-way.

Threshold quantity means the quantity specified for regulated substances pursuant to section 112(r)(5) of the Clean Air Act as amended, listed in §68.130 and determined to be present at a stationary source as specified in §68.115 of this part.

Typical meteorological conditions means the temperature, wind speed, cloud cover, and atmospheric stability class, prevailing at the site based on data gathered at or near the site or from a local meteorological station.

Vessel means any reactor, tank, drum, barrel, cylinder, vat, kettle, boiler, pipe, hose, or other container.

Worst-case release means the release of the largest quantity of a regulated substance from a vessel or process line failure that results in the greatest distance to an endpoint defined in §68.22(a).

[59 FR 4493, Jan. 31, 1994, as amended at 61 FR 31717, June 20, 1996; 63 FR 644, Jan. 6, 1998]

§68.10 Applicability.

(a) An owner or operator of a stationary source that has more than a threshold quantity of a regulated substance in a process, as determined under §68.115, shall comply with the requirements of this part no later than the latest of the following dates:

(1) June 21, 1999;

(2) Three years after the date on which a regulated substance is first listed under §68.130; or

(3) The date on which a regulated substance is first present above a threshold quantity in a process.

(b) Program 1 eligibility requirements. A covered process is eligible for Program 1 requirements as provided in §68.12(b) if it meets all of the following requirements:

(1) For the five years prior to the submission of an RMP, the process has not had an accidental release of a regulated substance where exposure to the substance, its reaction products, overpressure generated by an explosion involving the substance, or radiant heat generated by a fire involving the substance led to any of the following offsite:

(i) Death;

(ii) Injury; or

(iii) Response or restoration activities for an exposure of an environmental receptor;

(2) The distance to a toxic or flammable endpoint for a worst-case release assessment conducted under Subpart B and §68.25 is less than the distance to any public receptor, as defined in §68.30; and

(3) Emergency response procedures have been coordinated between the stationary source and local emergency planning and response organizations.

(c) Program 2 eligibility requirements. A covered process is subject to Program 2 requirements if it does not meet the eligibility requirements of either paragraph (b) or paragraph (d) of this section.

(d) Program 3 eligibility requirements. A covered process is subject to Program 3 if the process does not meet the requirements of paragraph (b) of this section, and if either of the following conditions is met:

(1) The process is in SIC code 2611, 2812, 2819, 2821, 2865, 2869, 2873, 2879, or 2911; or

(2) The process is subject to the OSHA process safety management standard, 29 CFR 1910.119.

(e) If at any time a covered process no longer meets the eligibility criteria of its Program level, the owner or operator shall comply with the requirements of the new Program level that applies to the process and update the RMP as provided in § 68.190.

(f) The provisions of this part shall not apply to an Outer Continental Shelf ("OCS") source, as defined in 40 CFR 55.2.

[61 FR 31717, June 20, 1996, as amended at 63 FR 645, Jan. 6, 1998]

§ 68.12 General requirements.

(a) General requirements. The owner or operator of a stationary source subject to this part shall submit a single RMP, as provided in §§ 68.150 to 68.185. The RMP shall include a registration that reflects all covered processes.

(b) Program 1 requirements. In addition to meeting the requirements of paragraph (a) of this section, the owner or operator of a stationary source with a process eligible for Program 1, as provided in § 68.10(b), shall:

(1) Analyze the worst-case release scenario for the process(es), as provided in §68.25; document that the nearest public receptor is beyond the distance to a toxic or flammable endpoint defined in §68.22(a); and submit in the RMP the worst-case release scenario as provided in §68.165;

(2) Complete the five-year accident history for the process as provided in §68.42 of this part and submit it in the RMP as provided in § 68.168;

(3) Ensure that response actions have been coordinated with local emergency planning and response agencies; and

(4) Certify in the RMP the following: "Based on the criteria in 40 CFR 68.10, the distance to the specified endpoint for the worst-case accidental release scenario for the following process(es) is less than the distance to the nearest public receptor: [list process(es)]. Within the past five years, the process(es) has (have) had no accidental release that caused offsite impacts provided in the risk management program rule (40 CFR 68.10(b)(1)). No additional measures are necessary to prevent offsite impacts from accidental releases. In the event of fire, explosion, or a release of a regulated substance from the process(es), entry within the distance to the specified endpoints may pose a danger to public emergency responders. Therefore, public emergency responders should not enter this area except as arranged with the emergency contact indicated in the RMP. The undersigned certifies that, to the best of my knowledge, information, and belief, formed after reasonable inquiry, the information submitted is true, accurate, and complete. [Signature, title, date signed]."

(c) Program 2 requirements. In addition to meeting the requirements of paragraph (a) of this section, the owner or operator of a stationary source with a process subject to Program 2, as provided in § 68.10(c), shall:

(1) Develop and implement a management system as provided in § 68.15;

(2) Conduct a hazard assessment as provided in §§ 68.20 through 68.42;

(3) Implement the Program 2 prevention steps provided in §§68.48 through 68.60 or implement the Program 3 prevention steps provided in §§ 68.65 through 68.87;

(4) Develop and implement an emergency response program as provided in §§ 68.90 to 68.95; and

(5) Submit as part of the RMP the data on prevention program elements for Program 2 processes as provided in §68.170.

(d) Program 3 requirements. In addition to meeting the requirements of paragraph (a) of this section, the owner or operator of a stationary source with a process subject to Program 3, as provided in § 68.10(d) shall:

(1) Develop and implement a management system as provided in § 68.15;

(2) Conduct a hazard assessment as provided in §§ 68.20 through 68.42;

(3) Implement the prevention requirements of §§ 68.65 through 68.87;

(4) Develop and implement an emergency response program as provided in §§ 68.90 to 68.95 of this part; and

(5) Submit as part of the RMP the data on prevention program elements

for Program 3 processes as provided in §68.175.

[61 FR 31718, June 20, 1996]

§ 68.15 Management.

(a) The owner or operator of a stationary source with processes subject to Program 2 or Program 3 shall develop a management system to oversee the implementation of the risk management program elements.

(b) The owner or operator shall assign a qualified person or position that has the overall responsibility for the development, implementation, and integration of the risk management program elements.

(c) When responsibility for implementing individual requirements of this part is assigned to persons other than the person identified under paragraph (b) of this section, the names or positions of these people shall be documented and the lines of authority defined through an organization chart or similar document.

[61 FR 31718, June 20, 1996]

Subpart B—Hazard Assessment

SOURCE: 61 FR 31718, June 20, 1996, unless otherwise noted.

§ 68.20 Applicability.

The owner or operator of a stationary source subject to this part shall prepare a worst-case release scenario analysis as provided in §68.25 of this part and complete the five-year accident history as provided in §68.42. The owner or operator of a Program 2 and 3 process must comply with all sections in this subpart for these processes.

§ 68.22 Offsite consequence analysis parameters.

(a) Endpoints. For analyses of offsite consequences, the following endpoints shall be used:

(1) Toxics. The toxic endpoints provided in appendix A of this part.

(2) Flammables. The endpoints for flammables vary according to the scenarios studied:

(i) Explosion. An overpressure of 1 psi.

(ii) Radiant heat/exposure time. A radiant heat of 5 kw/m^2 for 40 seconds.

(iii) Lower flammability limit. A lower flammability limit as provided in NFPA documents or other generally recognized sources.

(b) Wind speed/atmospheric stability class. For the worst-case release analysis, the owner or operator shall use a wind speed of 1.5 meters per second and F atmospheric stability class. If the owner or operator can demonstrate that local meteorological data applicable to the stationary source show a higher minimum wind speed or less stable atmosphere at all times during the previous three years, these minimums may be used. For analysis of alternative scenarios, the owner or operator may use the typical meteorological conditions for the stationary source.

(c) Ambient temperature/humidity. For worst-case release analysis of a regulated toxic substance, the owner or operator shall use the highest daily maximum temperature in the previous three years and average humidity for the site, based on temperature/humidity data gathered at the stationary source or at a local meteorological station; an owner or operator using the RMP Offsite Consequence Analysis Guidance may use 25°C and 50 percent humidity as values for these variables. For analysis of alternative scenarios, the owner or operator may use typical temperature/humidity data gathered at the stationary source or at a local meteorological station.

(d) Height of release. The worst-case release of a regulated toxic substance shall be analyzed assuming a ground level (0 feet) release. For an alternative scenario analysis of a regulated toxic substance, release height may be determined by the release scenario.

(e) Surface roughness. The owner or operator shall use either urban or rural topography, as appropriate. Urban means that there are many obstacles in the immediate area; obstacles include buildings or trees. Rural means there are no buildings in the immediate area and the terrain is generally flat and unobstructed.

(f) Dense or neutrally buoyant gases. The owner or operator shall ensure that tables or models used for dispersion analysis of regulated toxic substances appropriately account for gas density.

(g) Temperature of released substance. For worst case, liquids other than gases liquified by refrigeration only shall be considered to be released at the highest daily maximum temperature, based on data for the previous three years appropriate for the stationary source, or at process temperature, whichever is higher. For alternative scenarios, substances may be considered to be released at a process or ambient temperature that is appropriate for the scenario.

§ 68.25 Worst-case release scenario analysis.

(a) The owner or operator shall analyze and report in the RMP:

(1) For Program 1 processes, one worst-case release scenario for each Program 1 process;

(2) For Program 2 and 3 processes:

(i) One worst-case release scenario that is estimated to create the greatest distance in any direction to an endpoint provided in appendix A of this part resulting from an accidental release of regulated toxic substances from covered processes under worst-case conditions defined in § 68.22;

(ii) One worst-case release scenario that is estimated to create the greatest distance in any direction to an endpoint defined in § 68.22(a) resulting from an accidental release of regulated flammable substances from covered processes under worst-case conditions defined in § 68.22; and

(iii) Additional worst-case release scenarios for a hazard class if a worst-case release from another covered process at the stationary source potentially affects public receptors different from those potentially affected by the worst-case release scenario developed under paragraphs (a)(2)(i) or (a)(2)(ii) of this section.

(b) *Determination of worst-case release quantity.* The worst-case release quantity shall be the greater of the following:

(1) For substances in a vessel, the greatest amount held in a single vessel, taking into account administrative controls that limit the maximum quantity; or

(2) For substances in pipes, the greatest amount in a pipe, taking into account administrative controls that limit the maximum quantity.

(c) *Worst-case release scenario—toxic gases.* (1) For regulated toxic substances that are normally gases at ambient temperature and handled as a gas or as a liquid under pressure, the owner or operator shall assume that the quantity in the vessel or pipe, as determined under paragraph (b) of this section, is released as a gas over 10 minutes. The release rate shall be assumed to be the total quantity divided by 10 unless passive mitigation systems are in place.

(2) For gases handled as refrigerated liquids at ambient pressure:

(i) If the released substance is not contained by passive mitigation systems or if the contained pool would have a depth of 1 cm or less, the owner or operator shall assume that the substance is released as a gas in 10 minutes;

(ii) If the released substance is contained by passive mitigation systems in a pool with a depth greater than 1 cm, the owner or operator may assume that the quantity in the vessel or pipe, as determined under paragraph (b) of this section, is spilled instantaneously to form a liquid pool. The volatilization rate (release rate) shall be calculated at the boiling point of the substance and at the conditions specified in paragraph (d) of this section.

(d) *Worst-case release scenario—toxic liquids.* (1) For regulated toxic substances that are normally liquids at ambient temperature, the owner or operator shall assume that the quantity in the vessel or pipe, as determined under paragraph (b) of this section, is spilled instantaneously to form a liquid pool.

(i) The surface area of the pool shall be determined by assuming that the liquid spreads to 1 centimeter deep unless passive mitigation systems are in place that serve to contain the spill and limit the surface area. Where passive mitigation is in place, the surface area of the contained liquid shall be used to calculate the volatilization rate.

(ii) If the release would occur onto a surface that is not paved or smooth, the owner or operator may take into account the actual surface characteristics.

(2) The volatilization rate shall account for the highest daily maximum temperature occurring in the past three years, the temperature of the substance in the vessel, and the concentration of the substance if the liquid spilled is a mixture or solution.

(3) The rate of release to air shall be determined from the volatilization rate of the liquid pool. The owner or operator may use the methodology in the RMP Offsite Consequence Analysis Guidance or any other publicly available techniques that account for the modeling conditions and are recognized by industry as applicable as part of current practices. Proprietary models that account for the modeling conditions may be used provided the owner or operator allows the implementing agency access to the model and describes model features and differences from publicly available models to local emergency planners upon request.

(e) *Worst-case release scenario—flammables.* The owner or operator shall assume that the quantity of the substance, as determined under paragraph (b) of this section, vaporizes resulting in a vapor cloud explosion. A yield factor of 10 percent of the available energy released in the explosion shall be used to determine the distance to the explosion endpoint if the model used is based on TNT-equivalent methods.

(f) *Parameters to be applied.* The owner or operator shall use the parameters defined in § 68.22 to determine distance to the endpoints. The owner or operator may use the methodology provided in the RMP Offsite Consequence Analysis Guidance or any commercially or publicly available air dispersion modeling techniques, provided the techniques account for the modeling conditions and are recognized by industry as applicable as part of current practices. Proprietary models that account for the modeling conditions may be used provided the owner or operator allows the implementing agency access to the model and describes model features and differences from publicly available models to local emergency planners upon request.

(g) *Consideration of passive mitigation.* Passive mitigation systems may be considered for the analysis of worst case provided that the mitigation system is capable of withstanding the release event triggering the scenario and would still function as intended.

(h) *Factors in selecting a worst-case scenario.* Notwithstanding the provisions of paragraph (b) of this section, the owner or operator shall select as the worst case for flammable regulated substances or the worst case for regulated toxic substances, a scenario based on the following factors if such a scenario would result in a greater distance to an endpoint defined in § 68.22(a) beyond the stationary source boundary than the scenario provided under paragraph (b) of this section:

(1) Smaller quantities handled at higher process temperature or pressure; and

(2) Proximity to the boundary of the stationary source.

§ 68.28 Alternative release scenario analysis.

(a) The number of scenarios. The owner or operator shall identify and analyze at least one alternative release scenario for each regulated toxic substance held in a covered process(es) and at least one alternative release scenario to represent all flammable substances held in covered processes.

(b) *Scenarios to consider.* (1) For each scenario required under paragraph (a) of this section, the owner or operator shall select a scenario:

(i) That is more likely to occur than the worst-case release scenario under § 68.25; and

(ii) That will reach an endpoint offsite, unless no such scenario exists.

(2) Release scenarios considered should include, but are not limited to, the following, where applicable:

(i) Transfer hose releases due to splits or sudden hose uncoupling;

(ii) Process piping releases from failures at flanges, joints, welds, valves and valve seals, and drains or bleeds;

(iii) Process vessel or pump releases due to cracks, seal failure, or drain, bleed, or plug failure;

(iv) Vessel overfilling and spill, or overpressurization and venting through relief valves or rupture disks; and

(v) Shipping container mishandling and breakage or puncturing leading to a spill.

(c) Parameters to be applied. The owner or operator shall use the appropriate parameters defined in §68.22 to determine distance to the endpoints. The owner or operator may use either the methodology provided in the RMP Offsite Consequence Analysis Guidance or any commercially or publicly available air dispersion modeling techniques, provided the techniques account for the specified modeling conditions and are recognized by industry as applicable as part of current practices. Proprietary models that account for the modeling conditions may be used provided the owner or operator allows the implementing agency access to the model and describes model features and differences from publicly available models to local emergency planners upon request.

(d) Consideration of mitigation. Active and passive mitigation systems may be considered provided they are capable of withstanding the event that triggered the release and would still be functional.

(e) Factors in selecting scenarios. The owner or operator shall consider the following in selecting alternative release scenarios:

(1) The five-year accident history provided in §68.42; and

(2) Failure scenarios identified under §68.50 or §68.67.

§68.30 Defining offsite impacts—population.

(a) The owner or operator shall estimate in the RMP the population within a circle with its center at the point of the release and a radius determined by the distance to the endpoint defined in §68.22(a).

(b) *Population to be defined.* Population shall include residential population. The presence of institutions (schools, hospitals, prisons), parks and recreational areas, and major commercial, office, and industrial buildings shall be noted in the RMP.

(c) *Data sources acceptable.* The owner or operator may use the most recent Census data, or other updated information, to estimate the population potentially affected.

(d) *Level of accuracy.* Population shall be estimated to two significant digits.

§68.33 Defining offsite impacts—environment.

(a) The owner or operator shall list in the RMP environmental receptors within a circle with its center at the point of the release and a radius determined by the distance to the endpoint defined in §68.22(a) of this part.

(b) Data sources acceptable. The owner or operator may rely on information provided on local U.S. Geological Survey maps or on any data source containing U.S.G.S. data to identify environmental receptors.

68.36 Review and update.

(a) The owner or operator shall review and update the offsite consequence analyses at least once every five years.

(b) If changes in processes, quantities stored or handled, or any other aspect of the stationary source might reasonably be expected to increase or decrease the distance to the endpoint by a factor of two or more, the owner or operator shall complete a revised analysis within six months of the change and submit a revised risk management plan as provided in §68.190.

§68.39 Documentation.

The owner or operator shall maintain the following records on the offsite consequence analyses:

(a) For worst-case scenarios, a description of the vessel or pipeline and substance selected as worst case, assumptions and parameters used, and the rationale for selection; assumptions shall include use of any administrative controls and any passive mitigation that were assumed to limit the quantity that could be released. Documentation shall include the anticipated effect of the controls and mitigation on the release quantity and rate.

(b) For alternative release scenarios, a description of the scenarios identified, assumptions and parameters used, and the rationale for the selection of specific scenarios; assumptions shall include use of any administrative controls and any mitigation that were assumed to limit the quantity that could be released. Documentation shall include the effect of the controls and mitigation on the release quantity and rate.

(c) Documentation of estimated quantity released, release rate, and duration of release.

(d) Methodology used to determine distance to endpoints.

(e) Data used to estimate population and environmental receptors potentially affected.

§ 68.42 Five-year accident history.

(a) The owner or operator shall include in the five-year accident history all accidental releases from covered processes that resulted in deaths, injuries, or significant property damage on site, or known offsite deaths, injuries, evacuations, sheltering in place, property damage, or environmental damage.

(b) *Data required.* For each accidental release included, the owner or operator shall report the following information:

(1) Date, time, and approximate duration of the release;

(2) Chemical(s) released;

(3) Estimated quantity released in pounds;

(4) The type of release event and its source;

(5) Weather conditions, if known;

(6) On-site impacts;

(7) Known offsite impacts;

(8) Initiating event and contributing factors if known;

(9) Whether offsite responders were notified if known; and

(10) Operational or process changes that resulted from investigation of the release.

(c) *Level of accuracy.* Numerical estimates may be provided to two significant digits.

Subpart C—Program 2 Prevention Program

SOURCE: 61 FR 31721, June 20, 1996, unless otherwise noted.

§ 68.48 Safety information.

(a) The owner or operator shall compile and maintain the following up-to-date safety information related to the regulated substances, processes, and equipment:

(1) Material Safety Data Sheets that meet the requirements of 29 CFR 1910.1200(g);

(2) Maximum intended inventory of equipment in which the regulated substances are stored or processed;

(3) Safe upper and lower temperatures, pressures, flows, and compositions;

(4) Equipment specifications; and

(5) Codes and standards used to design, build, and operate the process.

(b) The owner or operator shall ensure that the process is designed in compliance with recognized and generally accepted good engineering practices. Compliance with Federal or state regulations that address industry-specific safe design or with industry-specific design codes and standards may be used to demonstrate compliance with this paragraph.

(c) The owner or operator shall update the safety information if a major change occurs that makes the information inaccurate.

§ 68.50 Hazard review.

(a) The owner or operator shall conduct a review of the hazards associated with the regulated substances, process, and procedures. The review shall identify the following:

(1) The hazards associated with the process and regulated substances;

(2) Opportunities for equipment malfunctions or human errors that could cause an accidental release;

(3) The safeguards used or needed to control the hazards or prevent equipment malfunction or human error; and

(4) Any steps used or needed to detect or monitor releases.

(b) The owner or operator may use checklists developed by persons or organizations knowledgeable about the process and equipment as a guide to conducting the review. For processes designed to meet industry standards or Federal or state design rules, the hazard review shall, by inspecting all equipment, determine whether the process is designed, fabricated, and operated in accordance with the applicable standards or rules.

(c) The owner or operator shall document the results of the review and ensure that problems identified are resolved in a timely manner.

(d) The review shall be updated at least once every five years. The owner or operator shall also conduct reviews whenever a major change in the process occurs; all issues identified in the review shall be resolved before startup of the changed process.

§ 68.52 Operating procedures.

(a) The owner or operator shall prepare written operating procedures that provide clear instructions or steps for safely conducting activities associated with each covered process consistent with the safety information for that process. Operating procedures or instructions provided by equipment manufacturers or developed by persons or organizations knowledgeable about the process and equipment may be used as a basis for a stationary source's operating procedures.

(b) The procedures shall address the following:

(1) Initial startup;

(2) Normal operations;

(3) Temporary operations;

(4) Emergency shutdown and operations;

(5) Normal shutdown;

(6) Startup following a normal or emergency shutdown or a major change that requires a hazard review;

(7) Consequences of deviations and steps required to correct or avoid deviations; and

(8) Equipment inspections.

(c) The owner or operator shall ensure that the operating procedures are updated, if necessary, whenever a major change occurs and prior to startup of the changed process.

§ 68.54 Training.

(a) The owner or operator shall ensure that each employee presently operating a process, and each employee newly assigned to a covered process have been trained or tested competent in the operating procedures provided in §68.52 that pertain to their duties. For those employees already operating a process on June 21, 1999, the owner or operator may certify in writing that the employee has the required knowledge, skills, and abilities to safely carry out the duties and responsibilities as provided in the operating procedures.

(b) Refresher training. Refresher training shall be provided at least every three years, and more often if necessary, to each employee operating a process to ensure that the employee understands and adheres to the current operating procedures of the process. The owner or operator, in consultation with the employees operating the process, shall determine the appropriate frequency of refresher training.

(c) The owner or operator may use training conducted under Federal or state regulations or under industry-specific standards or codes or training conducted by covered process equipment vendors to demonstrate compliance with this section to the extent that the training meets the requirements of this section.

(d) The owner or operator shall ensure that operators are trained in any updated or new procedures prior to startup of a process after a major change.

§ 68.56 Maintenance.

(a) The owner or operator shall prepare and implement procedures to maintain the on-going mechanical integrity of the process equipment. The owner or operator may use procedures or instructions provided by covered process equipment vendors or procedures in Federal or state regulations or industry codes as the basis for stationary source maintenance procedures.

(b) The owner or operator shall train or cause to be trained each employee involved in maintaining the on-going mechanical integrity of the process. To ensure that the employee can perform the job tasks in a safe manner, each such employee shall be trained in the hazards of the process, in how to avoid or correct unsafe conditions, and in the procedures applicable to the employee's job tasks.

(c) Any maintenance contractor shall ensure that each contract maintenance employee is trained to perform the maintenance procedures developed under paragraph (a) of this section.

(d) The owner or operator shall perform or cause to be performed inspections and tests on process equipment. Inspection and testing procedures shall follow recognized and generally accepted good engineering practices. The frequency of inspections and tests of process equipment shall be consistent with applicable manufacturers' recommendations, industry standards or codes, good engineering practices, and prior operating experience.

§ 68.58 Compliance audits.

(a) The owner or operator shall certify that they have evaluated compliance with the provisions of this subpart at least every three years to verify that the procedures and practices developed under the rule are adequate and are being followed.

(b) The compliance audit shall be conducted by at least one person knowledgeable in the process.

(c) The owner or operator shall develop a report of the audit findings.

(d) The owner or operator shall promptly determine and document an appropriate response to each of the findings of the compliance audit and document that deficiencies have been corrected.

(e) The owner or operator shall retain the two (2) most recent compliance audit reports. This requirement does not apply to any compliance audit report that is more than five years old.

§ 68.60 Incident investigation.

(a) The owner or operator shall investigate each incident which resulted in, or could reasonably have resulted in a catastrophic release.

(b) An incident investigation shall be initiated as promptly as possible, but not later than 48 hours following the incident.

(c) A summary shall be prepared at the conclusion of the investigation which includes at a minimum:

(1) Date of incident;

(2) Date investigation began;

(3) A description of the incident;

(4) The factors that contributed to the incident; and,

(5) Any recommendations resulting from the investigation.

(d) The owner or operator shall promptly address and resolve the investigation findings and recommendations. Resolutions and corrective actions shall be documented.

(e) The findings shall be reviewed with all affected personnel whose job tasks are affected by the findings.

(f) Investigation summaries shall be retained for five years.

Subpart D—Program 3 Prevention Program

SOURCE: 61 FR 31722, June 20, 1996, unless otherwise noted.

§ 68.65 Process safety information.

(a) In accordance with the schedule set forth in § 68.67, the owner or operator shall complete a compilation of written process safety information before conducting any process hazard analysis required by the rule. The compilation of written process safety information is to enable the owner or operator and the employees involved in operating the process to identify and understand the hazards posed by those processes involving regulated substances. This process safety information shall include information pertaining to the hazards of the regulated substances used or produced by the process, information pertaining to the technology of the process, and information pertaining to the equipment in the process.

(b) Information pertaining to the hazards of the regulated substances in the process. This information shall consist of at least the following:

(1) Toxicity information;

(2) Permissible exposure limits;

(3) Physical data;

(4) Reactivity data:

(5) Corrosivity data;

(6) Thermal and chemical stability data; and

(7) Hazardous effects of inadvertent mixing of different materials that could foreseeably occur.

NOTE TO PARAGRAPH (b): Material Safety Data Sheets meeting the requirements of 29 CFR 1910.1200(g) may be used to comply with this requirement to the extent they contain the information required by this subparagraph.

(c) Information pertaining to the technology of the process.

(1) Information concerning the technology of the process shall include at least the following:

(i) A block flow diagram or simplified process flow diagram;

(ii) Process chemistry;

(iii) Maximum intended inventory;

(iv) Safe upper and lower limits for such items as temperatures, pressures, flows or compositions; and,

(v) An evaluation of the consequences of deviations.

(2) Where the original technical information no longer exists, such information may be developed in conjunction with the process hazard analysis in sufficient detail to support the analysis.

(d) Information pertaining to the equipment in the process.

(1) Information pertaining to the equipment in the process shall include:

(i) Materials of construction;

(ii) Piping and instrument diagrams (P&ID's);

(iii) Electrical classification;

(iv) Relief system design and design basis;

(v) Ventilation system design;

(vi) Design codes and standards employed;

(vii) Material and energy balances for processes built after June 21, 1999; and

(viii) Safety systems (e.g. interlocks, detection or suppression systems).

(2) The owner or operator shall document that equipment complies with recognized and generally accepted good engineering practices.

(3) For existing equipment designed and constructed in accordance with codes, standards, or practices that are no longer in general use, the owner or operator shall determine and document that the equipment is designed, maintained, inspected, tested, and operating in a safe manner.

§68.67 Process hazard analysis.

(a) The owner or operator shall perform an initial process hazard analysis (hazard evaluation) on processes covered by this part. The process hazard analysis shall be appropriate to the complexity of the process and shall identify, evaluate, and control the hazards involved in the process. The owner or operator shall determine and document the priority order for conducting process hazard analyses based on a rationale which includes such considerations as extent of the process hazards, number of potentially affected employees, age of the process, and operating history of the process. The process hazard analysis shall be conducted as soon as possible, but not later than June 21, 1999. Process hazards analyses completed to comply with 29 CFR 1910.119(e) are acceptable as initial process hazards analyses. These process hazard analyses shall be updated and revalidated, based on their completion date.

(b) The owner or operator shall use one or more of the following methodologies that are appropriate to determine and evaluate the hazards of the process being analyzed.

(1) What-If;

(2) Checklist;

(3) What-If/Checklist;

(4) Hazard and Operability Study (HAZOP);

(5) Failure Mode and Effects Analysis (FMEA);

(6) Fault Tree Analysis; or

(7) An appropriate equivalent methodology.

(c) The process hazard analysis shall address:

(1) The hazards of the process;

(2) The identification of any previous incident which had a likely potential for catastrophic consequences.

(3) Engineering and administrative controls applicable to the hazards and their interrelationships such as appropriate application of detection methodologies to provide early warning of releases. (Acceptable detection methods might include process monitoring and control instrumentation with alarms, and detection hardware such as hydrocarbon sensors.);

(4) Consequences of failure of engineering and administrative controls;

(5) Stationary source siting;

(6) Human factors; and

(7) A qualitative evaluation of a range of the possible safety and health effects of failure of controls.

(d) The process hazard analysis shall be performed by a team with expertise in engineering and process operations, and the team shall include at least one employee who has experience and knowledge specific to the process being evaluated. Also, one member of the team must be knowledgeable in the specific process hazard analysis methodology being used.

(e) The owner or operator shall establish a system to promptly address the team's findings and recommendations; assure that the recommendations are resolved in a timely manner and that the resolution is documented; document what actions are to be taken; complete actions as soon as possible; develop a written schedule of when these actions are to be completed; communicate the actions to operating, maintenance and other employees whose work assignments are in the process and who may be affected by the recommendations or actions.

(f) At least every five (5) years after the completion of the initial process hazard analysis, the process hazard analysis shall be updated and revalidated by a team meeting the requirements in paragraph (d) of this section, to assure that the process hazard analysis is consistent with the current process. Updated and revalidated process hazard analyses completed to comply with 29 CFR 1910.119(e) are acceptable to meet the requirements of this paragraph.

(g) The owner or operator shall retain process hazards analyses and updates or revalidations for each process covered by this section, as well as the documented resolution of recommendations described in paragraph (e) of this section for the life of the process.

§ 68.69 Operating procedures.

(a) The owner or operator shall develop and implement written operating procedures that provide clear instructions for safely conducting activities involved in each covered process consistent with the process safety information and shall address at least the following elements.

(1) Steps for each operating phase:

(i) Initial startup;

(ii) Normal operations;

(iii) Temporary operations;

(iv) Emergency shutdown including the conditions under which emergency shutdown is required, and the assignment of shutdown responsibility to qualified operators to ensure that emergency shutdown is executed in a safe and timely manner.

(v) Emergency operations;

(vi) Normal shutdown; and,

(vii) Startup following a turnaround, or after an emergency shutdown.

(2) Operating limits:

(i) Consequences of deviation; and

(ii) Steps required to correct or avoid deviation.

(3) Safety and health considerations:

(i) Properties of, and hazards presented by, the chemicals used in the process;

(ii) Precautions necessary to prevent exposure, including engineering controls, administrative controls, and personal protective equipment;

(iii) Control measures to be taken if physical contact or airborne exposure occurs;

(iv) Quality control for raw materials and control of hazardous chemical inventory levels; and,

(v) Any special or unique hazards.

(4) Safety systems and their functions.

(b) Operating procedures shall be readily accessible to employees who work in or maintain a process.

(c) The operating procedures shall be reviewed as often as necessary to assure that they reflect current operating practice, including changes that result from changes in process chemicals, technology, and equipment, and changes to stationary sources. The owner or operator shall certify annually that these operating procedures are current and accurate.

(d) The owner or operator shall develop and implement safe work practices to provide for the control of hazards during operations such as lockout/tagout; confined space entry; opening process equipment or piping; and control over entrance into a stationary source by maintenance, contractor, laboratory, or other support personnel. These safe work practices shall apply to employees and contractor employees.

§ 68.71 Training.

(a) *Initial training.* (1) Each employee presently involved in operating a process, and each employee before being involved in operating a newly assigned process, shall be trained in an overview of the process and in the operating procedures as specified in §68.69. The training shall include emphasis on the specific safety and health hazards, emergency operations including shutdown, and safe work practices applicable to the employee's job tasks.

(2) In lieu of initial training for those employees already involved in operating a process on June 21, 1999 an owner or operator may certify in writing that the employee has the required knowledge, skills, and abilities to safely carry out the duties and responsibilities as specified in the operating procedures.

(b) *Refresher training.* Refresher training shall be provided at least every three years, and more often if necessary, to each employee involved in operating a process to assure that the employee understands and adheres to the current operating procedures of the process. The owner or operator, in consultation with the employees involved in operating the process, shall determine the appropriate frequency of refresher training.

(c) *Training documentation.* The owner or operator shall ascertain that each employee involved in operating a process has received and understood the training required by this paragraph. The owner or operator shall prepare a record which contains the identity of the employee, the date of training, and the means used to verify that the employee understood the training.

§68.73 Mechanical integrity.

(a) *Application.* Paragraphs (b) through (f) of this section apply to the following process equipment:

(1) Pressure vessels and storage tanks;

(2) Piping systems (including piping components such as valves);

(3) Relief and vent systems and devices;

(4) Emergency shutdown systems;

(5) Controls (including monitoring devices and sensors, alarms, and interlocks) and,

(6) Pumps.

(b) *Written procedures.* The owner or operator shall establish and implement written procedures to maintain the on-going integrity of process equipment.

(c) *Training for process maintenance activities.* The owner or operator shall train each employee involved in maintaining the on-going integrity of process equipment in an overview of that process and its hazards and in the procedures applicable to the employee's job tasks to assure that the employee can perform the job tasks in a safe manner.

(d) *Inspection and testing.* (1) Inspections and tests shall be performed on process equipment.

(2) Inspection and testing procedures shall follow recognized and generally accepted good engineering practices.

(3) The frequency of inspections and tests of process equipment shall be consistent with applicable manufacturers' recommendations and good engineering practices, and more frequently if determined to be necessary by prior operating experience.

(4) The owner or operator shall document each inspection and test that has been performed on process equipment. The documentation shall identify the date of the inspection or test, the name of the person who performed the inspection or test, the serial number or other identifier of the equipment on which the inspection or test was performed, a description of the inspection or test performed, and the results of the inspection or test.

(e) *Equipment deficiencies.* The owner or operator shall correct deficiencies in equipment that are outside acceptable limits (defined by the process safety information in §68.65) before further use or in a safe and timely manner when necessary means are taken to assure safe operation.

(f) *Quality assurance.* (1) In the construction of new plants and equipment, the owner or operator shall assure that equipment as it is fabricated is suitable for the process application for which they will be used.

(2) Appropriate checks and inspections shall be performed to assure that equipment is installed properly and consistent with design specifications and the manufacturer's instructions.

(3) The owner or operator shall assure that maintenance materials, spare parts and equipment are suitable for the process application for which they will be used.

§ 68.75 Management of change.

(a) The owner or operator shall establish and implement written procedures to manage changes (except for "replacements in kind") to process chemicals, technology, equipment, and procedures; and, changes to stationary sources that affect a covered process.

(b) The procedures shall assure that the following considerations are addressed prior to any change:

(1) The technical basis for the proposed change;

(2) Impact of change on safety and health;

(3) Modifications to operating procedures;

(4) Necessary time period for the change; and,

(5) Authorization requirements for the proposed change.

(c) Employees involved in operating a process and maintenance and contract employees whose job tasks will be affected by a change in the process shall be informed of, and trained in, the change prior to start-up of the process or affected part of the process.

(d) If a change covered by this paragraph results in a change in the process safety information required by § 68.65 of this part, such information shall be updated accordingly.

(e) If a change covered by this paragraph results in a change in the operating procedures or practices required by §68.69, such procedures or practices shall be updated accordingly.

§ 68.77 Pre-startup review.

(a) The owner or operator shall perform a pre-startup safety review for new stationary sources and for modified stationary sources when the modification is significant enough to require a change in the process safety information.

(b) The pre-startup safety review shall confirm that prior to the introduction of regulated substances to a process:

(1) Construction and equipment is in accordance with design specifications;

(2) Safety, operating, maintenance, and emergency procedures are in place and are adequate;

(3) For new stationary sources, a process hazard analysis has been performed and recommendations have been resolved or implemented before startup; and modified stationary sources meet the requirements contained in management of change, § 68.75.

(4) Training of each employee involved in operating a process has been completed.

§ 68.79 Compliance audits.

(a) The owner or operator shall certify that they have evaluated compliance with the provisions of this section at least every three years to verify that the procedures and practices developed under the standard are adequate and are being followed.

(b) The compliance audit shall be conducted by at least one person knowledgeable in the process.

(c) A report of the findings of the audit shall be developed.

(d) The owner or operator shall promptly determine and document an appropriate response to each of the findings of the compliance audit, and document that deficiencies have been corrected.

(e) The owner or operator shall retain the two (2) most recent compliance audit reports.

§ 68.81 Incident investigation.

(a) The owner or operator shall investigate each incident which resulted in, or could reasonably have resulted in a catastrophic release of a regulated substance.

(b) An incident investigation shall be initiated as promptly as possible, but not later than 48 hours following the incident.

(c) An incident investigation team shall be established and consist of at least one person knowledgeable in the process involved, including a contract employee if the incident involved work of the contractor, and other persons with appropriate knowledge and experience to thoroughly investigate and analyze the incident.

(d) A report shall be prepared at the conclusion of the investigation which includes at a minimum:

(1) Date of incident;

(2) Date investigation began;

(3) A description of the incident;

(4) The factors that contributed to the incident; and,

(5) Any recommendations resulting from the investigation.

(e) The owner or operator shall establish a system to promptly address and resolve the incident report findings and recommendations. Resolutions and corrective actions shall be documented.

(f) The report shall be reviewed with all affected personnel whose job tasks are relevant to the incident findings including contract employees where applicable.

(g) Incident investigation reports shall be retained for five years.

§ 68.83 Employee participation.

(a) The owner or operator shall develop a written plan of action regarding the implementation of the employee participation required by this section.

(b) The owner or operator shall consult with employees and their representatives on the conduct and development of process hazards analyses and on the development of the other elements of process safety management in this rule.

(c) The owner or operator shall provide to employees and their representatives access to process hazard analyses and to all other information required to be developed under this rule.

§ 68.85 Hot work permit.

(a) The owner or operator shall issue a hot work permit for hot work operations conducted on or near a covered process.

(b) The permit shall document that the fire prevention and protection requirements in 29 CFR 1910.252(a) have been implemented prior to beginning the hot work operations; it shall indicate the date(s) authorized for hot work; and identify the object on which hot work is to be performed. The permit shall be kept on file until completion of the hot work operations.

§ 68.87 Contractors.

(a) *Application.* This section applies to contractors performing maintenance or repair, turnaround, major renovation, or specialty work on or adjacent to a covered process. It does not apply to contractors providing incidental services which do not influence process safety, such as janitorial work, food and drink services, laundry, delivery or other supply services.

(b) *Owner or operator responsibilities.* (1) The owner or operator, when selecting a contractor, shall obtain and evaluate information regarding the contract owner or operator's safety performance and programs.

(2) The owner or operator shall inform contract owner or operator of the known potential fire, explosion, or toxic release hazards related to the contractor's work and the process.

(3) The owner or operator shall explain to the contract owner or operator the applicable provisions of subpart E of this part.

(4) The owner or operator shall develop and implement safe work practices consistent with § 68.69(d), to control the entrance, presence, and exit of the contract owner or operator and contract employees in covered process areas.

(5) The owner or operator shall periodically evaluate the performance of the contract owner or operator in fulfilling their obligations as specified in paragraph (c) of this section.

(c) *Contract owner or operator responsibilities.* (1) The contract owner or operator shall assure that each contract employee is trained in the work practices necessary to safely perform his/her job.

(2) The contract owner or operator shall assure that each contract employee is instructed in the known potential fire, explosion, or toxic release hazards related to his/her job and the process, and the applicable provisions of the emergency action plan.

(3) The contract owner or operator shall document that each contract employee has received and understood the training required by this section. The contract owner or operator shall prepare a record which contains the identity of the contract employee, the date of training, and the means used to verify that the employee understood the training.

(4) The contract owner or operator shall assure that each contract employee follows the safety rules of the stationary source including the safe work practices required by § 68.69(d).

(5) The contract owner or operator shall advise the owner or operator of

any unique hazards presented by the contract owner or operator's work, or of any hazards found by the contract owner or operator's work.

Subpart E—Emergency Response

SOURCE: 61 FR 31725, June 20, 1996, unless otherwise noted.

§ 68.90 Applicability.

(a) Except as provided in paragraph (b) of this section, the owner or operator of a stationary source with Program 2 and Program 3 processes shall comply with the requirements of § 68.95.

(b) The owner or operator of stationary source whose employees will not respond to accidental releases of regulated substances need not comply with §68.95 of this part provided that they meet the following:

(1) For stationary sources with any regulated toxic substance held in a process above the threshold quantity, the stationary source is included in the community emergency response plan developed under 42 U.S.C. 11003;

(2) For stationary sources with only regulated flammable substances held in a process above the threshold quantity, the owner or operator has coordinated response actions with the local fire department; and

(3) Appropriate mechanisms are in place to notify emergency responders when there is a need for a response.

§ 68.95 Emergency response program.

(a) The owner or operator shall develop and implement an emergency response program for the purpose of protecting public health and the environment. Such program shall include the following elements:

(1) An emergency response plan, which shall be maintained at the stationary source and contain at least the following elements:

(i) Procedures for informing the public and local emergency response agencies about accidental releases;

(ii) Documentation of proper first-aid and emergency medical treatment necessary to treat accidental human exposures; and

(iii) Procedures and measures for emergency response after an accidental release of a regulated substance;

(2) Procedures for the use of emergency response equipment and for its inspection, testing, and maintenance;

(3) Training for all employees in relevant procedures; and

(4) Procedures to review and update, as appropriate, the emergency response plan to reflect changes at the stationary source and ensure that employees are informed of changes.

(b) A written plan that complies with other Federal contingency plan regulations or is consistent with the approach in the National Response Team's Integrated Contingency Plan Guidance ("One Plan") and that, among other matters, includes the elements provided in paragraph (a) of this section, shall satisfy the requirements of this section if the owner or operator also complies with paragraph (c) of this section.

(c) The emergency response plan developed under paragraph (a)(1) of this section shall be coordinated with the community emergency response plan developed under 42 U.S.C. 11003. Upon request of the local emergency planning committee or emergency response officials, the owner or operator shall promptly provide to the local emergency response officials information necessary for developing and implementing the community emergency response plan.

Subpart F—Regulated Substances for Accidental Release Prevention

SOURCE: 59 FR 4493, Jan. 31, 1994, unless otherwise noted. Redesignated at 61 FR 31717, June 20, 1996.

§ 68.100 Purpose.

This subpart designates substances to be listed under section 112(r)(3), (4), and (5) of the Clean Air Act, as amended, identifies their threshold quantities, and establishes the requirements for petitioning to add or delete substances from the list.

§ 68.115 Threshold determination.

(a) A threshold quantity of a regulated substance listed in § 68.130 is present at a stationary source if the total quantity of the regulated substance contained in a process exceeds the threshold.

(b) For the purposes of determining whether more than a threshold quantity of a regulated substance is present at the stationary source, the following exemptions apply:

(1) *Concentrations of a regulated toxic substance in a mixture.* If a regulated substance is present in a mixture and the concentration of the substance is below one percent by weight of the mixture, the amount of the substance in the mixture need not be considered when determining whether more than a threshold quantity is present at the stationary source. Except for oleum, toluene 2,4-diisocyanate, toluene 2,6-diisocyanate, and toluene diisocyanate (unspecified isomer), if the concentration of the regulated substance in the mixture is one percent or greater by weight, but the owner or operator can demonstrate that the partial pressure of the regulated substance in the mixture (solution) under handling or storage conditions in any portion of the process is less than 10 millimeters of mercury (mm Hg), the amount of the substance in the mixture in that portion of the process need not be considered when determining whether more than a threshold quantity is present at the stationary source. The owner or operator shall document this partial pressure measurement or estimate.

(2) *Concentrations of a regulated flammable substance in a mixture.* (i) *General provision.* If a regulated substance is present in a mixture and the concentration of the substance is below one percent by weight of the mixture, the mixture need not be considered when determining whether more than a threshold quantity of the regulated substance is present at the stationary source. Except as provided in paragraph (b)(2) (ii) and (iii) of this section, if the concentration of the substance is one percent or greater by weight of the mixture, then, for purposes of determining whether a threshold quantity is present at the stationary source, the entire weight of the mixture shall be treated as the regulated substance unless the owner or operator can demonstrate that the mixture itself does not have a National Fire Protection Association flammability hazard rating of 4. The demonstration shall be in accordance with the definition of flammability hazard rating 4 in the NFPA 704, Standard System for the Identification of the Hazards of Materials for Emergency Response, National Fire Protection Association, Quincy, MA, 1996. Available from the National Fire Protection Association, 1 Batterymarch Park, Quincy, MA 02269–9101. This incorporation by reference was approved by the Director of the Federal Register in accordance with 5 U.S.C. 552(a) and 1 CFR part 51. Copies may be inspected at the Environmental Protection Agency Air Docket (6102), Attn: Docket No. A–96–O8, Waterside Mall, 401 M. St. SW., Washington DC; or at the Office of Federal Register at 800 North Capitol St., NW, Suite 700, Washington, DC. Boiling point and flash point shall be defined and determined in accordance with NFPA 30, Flammable and Combustible Liquids Code, National Fire Protection Association, Quincy, MA, 1996. Available from the National Fire Protection Association, 1 Batterymarch Park, Quincy, MA 02269–9101. This incorporation by reference was approved by the Director of the Federal Register in accordance with 5 U.S.C. 552(a) and 1 CFR part 51. Copies may be inspected at the Environmental Protection Agency Air Docket (6102), Attn: Docket No. A–96–O8, Waterside Mall, 401 M. St. SW., Washington DC; or at the Office of Federal Register at 800 North Capitol St., NW., Suite 700, Washington, DC. The owner or operator shall document the National Fire Protection Association flammability hazard rating.

(ii) *Gasoline.* Regulated substances in gasoline, when in distribution or related storage for use as fuel for internal combustion engines, need not be considered when determining whether more than a threshold quantity is present at a stationary source.

(iii) *Naturally occurring hydrocarbon mixtures.* Prior to entry into a natural gas processing plant or a petroleum refining process unit, regulated substances in naturally occurring hydrocarbon mixtures need not be considered when determining whether more than a threshold quantity is present at a stationary source. Naturally occurring hydrocarbon mixtures include any combination of the following: condensate, crude oil, field gas, and produced water, each as defined in § 68.3 of this part.

(3) *Articles*. Regulated substances contained in articles need not be considered when determining whether more than a threshold quantity is present at the stationary source.

(4) *Uses*. Regulated substances, when in use for the following purposes, need not be included in determining whether more than a threshold quantity is present at the stationary source:

(i) Use as a structural component of the stationary source;

(ii) Use of products for routine janitorial maintenance;

(iii) Use by employees of foods, drugs, cosmetics, or other personal items containing the regulated substance; and

(iv) Use of regulated substances present in process water or non-contact cooling water as drawn from the environment or municipal sources, or use of regulated substances present in air used either as compressed air or as part of combustion.

(5) *Activities in laboratories*. If a regulated substance is manufactured, processed, or used in a laboratory at a stationary source under the supervision of a technically qualified individual as defined in § 720.3(ee) of this chapter, the quantity of the substance need not be considered in determining whether a threshold quantity is present. This exemption does not apply to:

(i) Specialty chemical production;

(ii) Manufacture, processing, or use of substances in pilot plant scale operations; and

(iii) Activities conducted outside the laboratory.

[59 FR 4493, Jan. 31, 1994. Redesignated at 61 FR 31717, June 20, 1996, as amended at 63 FR 645, Jan. 6, 1998]

§ 68.120 Petition process.

(a) Any person may petition the Administrator to modify, by addition or deletion, the list of regulated substances identified in § 68.130. Based on the information presented by the petitioner, the Administrator may grant or deny a petition.

(b) A substance may be added to the list if, in the case of an accidental release, it is known to cause or may be reasonably anticipated to cause death, injury, or serious adverse effects to human health or the environment.

(c) A substance may be deleted from the list if adequate data on the health and environmental effects of the substance are available to determine that the substance, in the case of an accidental release, is not known to cause and may not be reasonably anticipated to cause death, injury, or serious adverse effects to human health or the environment.

(d) No substance for which a national primary ambient air quality standard has been established shall be added to the list. No substance regulated under title VI of the Clean Air Act, as amended, shall be added to the list.

(e) The burden of proof is on the petitioner to demonstrate that the criteria for addition and deletion are met. A petition will be denied if this demonstration is not made.

(f) The Administrator will not accept additional petitions on the same substance following publication of a final notice of the decision to grant or deny a petition, unless new data becomes available that could significantly affect the basis for the decision.

(g) Petitions to modify the list of regulated substances must contain the following:

(1) Name and address of the petitioner and a brief description of the organization(s) that the petitioner represents, if applicable;

(2) Name, address, and telephone number of a contact person for the petition;

(3) Common chemical name(s), common synonym(s), Chemical Abstracts Service number, and chemical formula and structure;

(4) Action requested (add or delete a substance);

(5) Rationale supporting the petitioner's position; that is, how the substance meets the criteria for addition and deletion. A short summary of the rationale must be submitted along with a more detailed narrative; and

(6) Supporting data; that is, the petition must include sufficient information to scientifically support the request to modify the list. Such information shall include:

(i) A list of all support documents;

(ii) Documentation of literature searches conducted, including, but not limited to, identification of the database(s) searched, the search strategy, dates covered, and printed results;

(iii) Effects data (animal, human, and environmental test data) indicating the potential for death, injury, or serious adverse human and environmental impacts from acute exposure following an accidental release; printed copies of the data sources, in English, should be provided; and

(iv) Exposure data or previous accident history data, indicating the potential for serious adverse human health or environmental effects from an accidental release. These data may include, but are not limited to, physical and chemical properties of the substance, such as vapor pressure; modeling results, including data and assumptions used and model documentation; and historical accident data, citing data sources.

(h) Within 18 months of receipt of a petition, the Administrator shall publish in the FEDERAL REGISTER a notice either denying the petition or granting the petition and proposing a listing.

§ 68.125 Exemptions.

Agricultural nutrients. Ammonia used as an agricultural nutrient, when held by farmers, is exempt from all provisions of this part.

§ 68.130 List of substances.

(a) Regulated toxic and flammable substances under section 112(r) of the Clean Air Act are the substances listed in Tables 1, 2, 3, and 4. Threshold quantities for listed toxic and flammable substances are specified in the tables.

(b) The basis for placing toxic and flammable substances on the list of regulated substances are explained in the notes to the list.

TABLE 1 TO § 68.130.—LIST OF REGULATED TOXIC SUBSTANCES AND THRESHOLD QUANTITIES FOR ACCIDENTAL RELEASE PREVENTION

[Alphabetical Order—77 Substances]

Chemical name	CAS No.	Threshold quantity (lbs)	Basis for listing
Acrolein [2-Propenal]	107–02–8	5,000	b
Acrylonitrile [2-Propenenitrile]	107–13–1	20,000	b
Acrylyl chloride [2-Propenoyl chloride]	814–68–6	5,000	b
Allyl alcohol [2-Propen-1-ol]	107–18–61	15,000	b
Allylamine [2-Propen-1-amine]	107–11–9	10,000	b
Ammonia (anhydrous)	7664–41–7	10,000	a, b
Ammonia (conc 20% or greater)	7664–41–7	20,000	a, b
Arsenous trichloride	7784–34–1	15,000	b
Arsine	7784–42–1	1,000	b
Boron trichloride [Borane, trichloro-]	10294–34–5	5,000	b
Boron trifluoride [Borane, trifluoro-]	7637–07–2	5,000	b
Boron trifluoride compound with methyl ether (1:1) [Boron, trifluoro [oxybis [metane]]-, T-4-	353–42–4	15,000	b
Bromine	7726–95–6	10,000	a, b
Carbon disulfide	75–15–0	20,000	b
Chlorine	7782–50–5	2,500	a, b
Chlorine dioxide [Chlorine oxide (ClO2)]	10049–04–4	1,000	c
Chloroform [Methane, trichloro-]	67–66–3	20,000	b
Chloromethyl ether [Methane, oxybis[chloro-]	542–88–1	1,000	b
Chloromethyl methyl ether [Methane, chloromethoxy-]	107–30–2	5,000	b
Crotonaldehyde [2-Butenal]	4170–30–3	20,000	b
Crotonaldehyde, (E)- [2-Butenal, (E)-]	123–73–9	20,000	b
Cyanogen chloride	506–77–4	10,000	c
Cyclohexylamine [Cyclohexanamine]	108–91–8	15,000	b
Diborane	19287–45–7	2,500	b
Dimethyldichlorosilane [Silane, dichlorodimethyl-]	75–78–5	5,000	b
1,1-Dimethylhydrazine [Hydrazine, 1,1-dimethyl-]	57–14–7	15,000	b

TABLE 1 TO § 68.130.—LIST OF REGULATED TOXIC SUBSTANCES AND THRESHOLD QUANTITIES FOR ACCIDENTAL RELEASE PREVENTION—Continued

[Alphabetical Order—77 Substances]

Chemical name	CAS No.	Threshold quantity (lbs)	Basis for listing
Epichlorohydrin [Oxirane, (chloromethyl)-]	106–89–8	20,000	b
Ethylenediamine [1,2-Ethanediamine]	107–15–3	20,000	b
Ethyleneimine [Aziridine]	151–56–4	10,000	b
Ethylene oxide [Oxirane]	75–21–8	10,000	a, b
Fluorine	7782–41–4	1,000	b
Formaldehyde (solution)	50–00–0	15,000	b
Furan	110–00–9	5,000	b
Hydrazine	302–01–2	15,000	b
Hydrochloric acid (conc 37% or greater)	7647–01–0	15,000	d
Hydrocyanic acid	74–90–8	2,500	a, b
Hydrogen chloride (anhydrous) [Hydrochloric acid]	7647–01–0	5,000	a
Hydrogen fluoride/Hydrofluoric acid (conc 50% or greater) [Hydrofluoric acid]	7664–39–3	1,000	a, b
Hydrogen selenide	7783–07–5	500	b
Hydrogen sulfide	7783–06–4	10,000	a, b
Iron, pentacarbonyl- [Iron carbonyl (Fe(CO)5), (TB-5-11)-]	13463–40–6	2,500	b
Isobutyronitrile [Propanenitrile, 2-methyl-]	78–82–0	20,000	b
Isopropyl chloroformate [Carbonochloridic acid, 1-methylethyl ester]	108–23–6	15,000	b
Methacrylonitrile [2-Propenenitrile, 2-methyl-]	126–98–7	10,000	b
Methyl chloride [Methane, chloro-]	74–87–3	10,000	a
Methyl chloroformate [Carbonochloridic acid, methylester]	79–22–1	5,000	b
Methyl hydrazine [Hydrazine, methyl-]	60–34–4	15,000	b
Methyl isocyanate [Methane, isocyanato-]	624–83–9	10,000	a, b
Methyl mercaptan [Methanethiol]	74–93–1	10,000	b
Methyl thiocyanate [Thiocyanic acid, methyl ester]	556–64–9	20,000	b
Methyltrichlorosilane [Silane, trichloromethyl-]	75–79–6	5,000	b
Nickel carbonyl	13463–39–3	1,000	b
Nitric acid (conc 80% or greater)	7697–37–2	15,000	b
Nitric oxide [Nitrogen oxide (NO)]	10102–43–9	10,000	b
Oleum (Fuming Sulfuric acid) [Sulfuric acid, mixture with sulfur trioxide][1]	8014–95–7	10,000	e
Peracetic acid [Ethaneperoxoic acid]	79–21–0	10,000	b
Perchloromethylmercaptan [Methanesulfenyl chloride, trichloro-]	594–42–3	10,000	b
Phosgene [Carbonic dichloride]	75–44–5	500	a, b
Phosphine	7803–51–2	5,000	b
Phosphorus oxychloride [Phosphoryl chloride]	10025–87–3	5,000	b
Phosphorus trichloride [Phosphorous trichloride]	7719–12–2	15,000	b
Piperidine	110–89–4	15,000	b
Propionitrile [Propanenitrile]	107–12–0	10,000	b
Propyl chloroformate [Carbonochloridic acid, propylester]	109–61–5	15,000	b
Propyleneimine [Aziridine, 2-methyl-]	75–55–8	10,000	b
Propylene oxide [Oxirane, methyl-]	75–56–9	10,000	b
Sulfur dioxide (anhydrous)	7446–09–5	5,000	a, b
Sulfur tetrafluoride [Sulfur fluoride (SF4), (T-4)-]	7783–60–0	2,500	b
Sulfur trioxide	7446–11–9	10,000	a, b
Tetramethyllead [Plumbane, tetramethyl-]	75–74–1	10,000	b
Tetranitromethane [Methane, tetranitro-]	509–14–8	10,000	b

TABLE 1 TO § 68.130.—LIST OF REGULATED TOXIC SUBSTANCES AND THRESHOLD QUANTITIES FOR ACCIDENTAL RELEASE PREVENTION—Continued

[Alphabetical Order—77 Substances]

Chemical name	CAS No.	Threshold quantity (lbs)	Basis for listing
Titanium tetrachloride [Titanium chloride (TiCl4) (T-4)-]	7550–45–0	2,500	b
Toluene 2,4-diisocyanate [Benzene, 2,4-diisocyanato-1-methyl-][1]	584–84–9	10,000	a
Toluene 2,6-diisocyanate [Benzene, 1,3-diisocyanato-2-methyl-][1]	91–08–7	10,000	a
Toluene diisocyanate (unspecified isomer) [Benzene, 1,3-diisocyanatomethyl-][1].	26471–62–5	10,000	a
Trimethylchlorosilane [Silane, chlorotrimethyl-]	75–77–4	10,000	b
Vinyl acetate monomer [Acetic acid ethenyl ester]	108–05–4	15,000	b

[1] The mixture exemption in § 68.115(b)(1) does not apply to the substance.

NOTE: Basis for Listing:

a Mandated for listing by Congress.

b On EHS list, vapor pressure 10 mmHg or greater.

c Toxic gas.

d Toxicity of hydrogen chloride, potential to release hydrogen chloride, and history of accidents.

e Toxicity of sulfur trioxide and sulfuric acid, potential to release sulfur trioxide, and history of accidents.

TABLE 2 TO § 68.130.—LIST OF REGULATED TOXIC SUBSTANCES AND THRESHOLD QUANTITIES FOR ACCIDENTAL RELEASE PREVENTION

[CAS Number Order—77 Substances]

CAS No.	Chemical name	Threshold quantity (lbs)	Basis for listing
50–00–0	Formaldehyde (solution)	15,000	b
57–14–7	1,1-Dimethylhydrazine [Hydrazine, 1,1-dimethyl-]	15,000	b
60–34–4	Methyl hydrazine [Hydrazine, methyl-]	15,000	b
67–66–3	Chloroform [Methane, trichloro-]	20,000	b
74–87–3	Methyl chloride [Methane, chloro-]	10,000	a
74–90–8	Hydrocyanic acid	2,500	a, b
74–93–1	Methyl mercaptan [Methanethiol]	10,000	b
75–15–0	Carbon disulfide	20,000	b
75–21–8	Ethylene oxide [Oxirane]	10,000	a, b
75–44–5	Phosgene [Carbonic dichloride]	500	a, b
75–55–8	Propyleneimine [Aziridine, 2-methyl-]	10,000	b
75–56–9	Propylene oxide [Oxirane, methyl-]	10,000	b
75–74–1	Tetramethyllead [Plumbane, tetramethyl-]	10,000	b
75–77–4	Trimethylchlorosilane [Silane, chlorotrimethyl-]	10,000	b
75–78–5	Dimethyldichlorosilane [Silane, dichlorodimethyl-]	5,000	b
75–79–6	Methyltrichlorosilane [Silane, trichloromethyl-]	5,000	b
78–82–0	Isobutyronitrile [Propanenitrile, 2-methyl-]	20,000	b
79–21–0	Peracetic acid [Ethaneperoxoic acid]	10,000	b
79–22–1	Methyl chloroformate [Carbonochloridic acid, methylester]	5,000	b
91–08–7	Toluene 2,6-diisocyanate [Benzene, 1,3-diisocyanato-2-methyl-][1]	10,000	a
106–89–8	Epichlorohydrin [Oxirane, (chloromethyl)-]	20,000	b
107–02–8	Acrolein [2-Propenal]	5,000	b
107–11–9	Allylamine [2-Propen-1-amine]	10,000	b
107–12–0	Propionitrile [Propanenitrile]	10,000	b
107–13–1	Acrylonitrile [2-Propenenitrile]	20,000	b
107–15–3	Ethylenediamine [1,2-Ethanediamine]	20,000	b
107–18–6	Allyl alcohol [2-Propen-1-ol]	15,000	b
107–30–2	Chloromethyl methyl ether [Methane, chloromethoxy-]	5,000	b
108–05–4	Vinyl acetate monomer [Acetic acid ethenyl ester]	15,000	b
108–23–6	Isopropyl chloroformate [Carbonochloridic acid, 1-methylethyl ester]	15,000	b
108–91–8	Cyclohexylamine [Cyclohexanamine]	15,000	b
109–61–5	Propyl chloroformate [Carbonochloridic acid, propylester]	15,000	b
110–00–9	Furan	5,000	b
110–89–4	Piperidine	15,000	b
123–73–9	Crotonaldehyde, (E)- [2-Butenal, (E)-]	20,000	b

TABLE 2 TO § 68.130.—LIST OF REGULATED TOXIC SUBSTANCES AND THRESHOLD QUANTITIES FOR ACCIDENTAL RELEASE PREVENTION—Continued

[CAS Number Order—77 Substances]

CAS No.	Chemical name	Threshold quantity (lbs)	Basis for listing
126–98–7	Methacrylonitrile [2-Propenenitrile, 2-methyl-]	10,000	b
151–56–4	Ethyleneimine [Aziridine]	10,000	b
302–01–2	Hydrazine	15,000	b
353–42–4	Boron trifluoride compound with methyl ether (1:1) [Boron, trifluoro[oxybis[methane]]-, T-4-	15,000	b
506–77–4	Cyanogen chloride	10,000	c
509–14–8	Tetranitromethane [Methane, tetranitro-]	10,000	b
542–88–1	Chloromethyl ether [Methane, oxybis[chloro-]	1,000	b
556–64–9	Methyl thiocyanate [Thiocyanic acid, methyl ester]	20,000	b
584–84–9	Toluene 2,4-diisocyanate [Benzene, 2,4-diisocyanato-1-methyl-][1]	10,000	a
594–42–3	Perchloromethylmercaptan [Methanesulfenyl chloride, trichloro-]	10,000	b
624–83–9	Methyl isocyanate [Methane, isocyanato-]	10,000	a, b
814–68–6	Acrylyl chloride [2-Propenoyl chloride]	5,000	b
4170–30–3	Crotonaldehyde [2-Butenal]	20,000	b
7446–09–5	Sulfur dioxide (anhydrous)	5,000	a, b
7446–11–9	Sulfur trioxide	10,000	a, b
7550–45–0	Titanium tetrachloride [Titanium chloride (TiCl4) (T-4)-]	2,500	b
7637–07–2	Boron trifluoride [Borane, trifluoro-]	5,000	b
7647–01–0	Hydrochloric acid (conc 37% or greater)	15,000	d
7647–01–0	Hydrogen chloride (anhydrous) [Hydrochloric acid]	5,000	a
7664–39–3	Hydrogen fluoride/Hydrofluoric acid (conc 50% or greater) [Hydrofluoric acid]	1,000	a, b
7664–41–7	Ammonia (anhydrous)	10,000	a, b
7664–41–7	Ammonia (conc 20% or greater)	20,000	a, b
7697–37–2	Nitric acid (conc 80% or greater)	15,000	b
7719–12–2	Phosphorus trichloride [Phosphorous trichloride]	15,000	b
7726–95–6	Bromine	10,000	a, b
7782–41–4	Fluorine	1,000	b
7782–50–5	Chlorine	2,500	a, b
7783–06–4	Hydrogen sulfide	10,000	a, b
7783–07–5	Hydrogen selenide	500	b
7783–60–0	Sulfur tetrafluoride [Sulfur fluoride (SF4), (T-4)-]	2,500	b
7784–34–1	Arsenous trichloride	15,000	b
7784–42–1	Arsine	1,000	b
7803–51–2	Phosphine	5,000	b
8014–95–7	Oleum (Fuming Sulfuric acid) [Sulfuric acid, mixture with sulfur trioxide][1]	10,000	e
10025–87–3	Phosphorus oxychloride [Phosphoryl chloride]	5,000	b
10049–04–4	Chlorine dioxide [Chlorine oxide (ClO_2)]	1,000	c
10102–43–9	Nitric oxide [Nitrogen oxide (NO)]	10,000	b
10294–34–5	Boron trichloride [Borane, trichloro-]	5,000	b
13463–39–3	Nickel carbonyl	1,000	b
13463–40–6	Iron, pentacarbonyl- [Iron carbonyl ($Fe(CO)_5$), (TB-5-11)-]	2,500	b
19287–45–7	Diborane	2,500	b
26471–62–5	Toluene diisocyanate (unspecified isomer) [Benzene, 1,3-diisocyanatomethyl-][1]	10,000	a

[1] The mixture exemption in § 68.115(b)(1) does not apply to the substance.

NOTE: Basis for Listing:

a Mandated for listing by Congress.

b On EHS list, vapor pressure 10 mmHg or greater.

c Toxic gas.

d Toxicity of hydrogen chloride, potential to release hydrogen chloride, and history of accidents.

e Toxicity of sulfur trioxide and sulfuric acid, potential to release sulfur trioxide, and history of accidents.

TABLE 3 TO § 68.130.—LIST OF REGULATED FLAMMABLE SUBSTANCES AND THRESHOLD QUANTITIES FOR ACCIDENTAL RELEASE PREVENTION

[Alphabetical Order—63 Substances]

Chemical name	CAS No.	Threshold quantity (lbs)	Basis for listing
Acetaldehyde	75–07–0	10,000	g
Acetylene [Ethyne]	74–86–2	10,000	f
Bromotrifluorethylene [Ethene, bromotrifluoro-]	598–73–2	10,000	f
1,3-Butadiene	106–99–0	10,000	f
Butane	106–97–8	10,000	f
1-Butene	106–98–9	10,000	f
2-Butene	107–01–7	10,000	f
Butene	25167–67–3	10,000	f

TABLE 3 TO § 68.130.—LIST OF REGULATED FLAMMABLE SUBSTANCES AND THRESHOLD QUANTITIES FOR ACCIDENTAL RELEASE PREVENTION—Continued

[Alphabetical Order—63 Substances]

Chemical name	CAS No.	Threshold quantity (lbs)	Basis for listing
2-Butene-cis	590–18–1	10,000	f
2-Butene-trans [2-Butene, (E)]	624–64–6	10,000	f
Carbon oxysulfide [Carbon oxide sulfide (COS)]	463–58–1	10,000	f
Chlorine monoxide [Chlorine oxide]	7791–21–1	10,000	f
2-Chloropropylene [1-Propene, 2-chloro-]	557–98–2	10,000	g
1-Chloropropylene [1-Propene, 1-chloro-]	590–21–6	10,000	g
Cyanogen [Ethanedinitrile]	460–19–5	10,000	f
Cyclopropane	75–19–4	10,000	f
Dichlorosilane [Silane, dichloro-]	4109–96–0	10,000	f
Difluoroethane [Ethane, 1,1-difluoro-]	75–37–6	10,000	f
Dimethylamine [Methanamine, N-methyl-]	124–40–3	10,000	f
2,2-Dimethylpropane [Propane, 2,2-dimethyl-]	463–82–1	10,000	f
Ethane	74–84–0	10,000	f
Ethyl acetylene [1-Butyne]	107–00–6	10,000	f
Ethylamine [Ethanamine]	75–04–7	10,000	f
Ethyl chloride [Ethane, chloro-]	75–00–3	10,000	f
Ethylene [Ethene]	74–85–1	10,000	f
Ethyl ether [Ethane, 1,1'-oxybis-]	60–29–7	10,000	g
Ethyl mercaptan [Ethanethiol]	75–08–1	10,000	g
Ethyl nitrite [Nitrous acid, ethyl ester]	109–95–5	10,000	f
Hydrogen	1333–74–0	10,000	f
Isobutane [Propane, 2-methyl]	75–28–5	10,000	f
Isopentane [Butane, 2-methyl-]	78–78–4	10,000	g
Isoprene [1,3-Butadinene, 2-methyl-]	78–79–5	10,000	g
Isopropylamine [2-Propanamine]	75–31–0	10,000	g
Isopropyl chloride [Propane, 2-chloro-]	75–29–6	10,000	g
Methane	74–82–8	10,000	f
Methylamine [Methanamine]	74–89–5	10,000	f
3-Methyl-1-butene	563–45–1	10,000	f
2-Methyl-1-butene	563–46–2	10,000	g
Methyl ether [Methane, oxybis-]	115–10–6	10,000	f
Methyl formate [Formic acid, methyl ester]	107–31–3	10,000	g
2-Methylpropene [1-Propene, 2-methyl-]	115–11–7	10,000	f
1,3-Pentadinene	504–60–9	10,000	f
Pentane	109–66–0	10,000	g
1-Pentene	109–67–1	10,000	g
2-Pentene, (E)-	646–04–8	10,000	g
2-Pentene, (Z)-	627–20–3	10,000	g
Propadiene [1,2-Propadiene]	463–49–0	10,000	f
Propane	74–98–6	10,000	f
Propylene [1-Propene]	115–07–1	10,000	f
Propyne [1-Propyne]	74–99–7	10,000	f
Silane	7803–62–5	10,000	f
Tetrafluoroethylene [Ethene, tetrafluoro-]	116–14–3	10,000	f
Tetramethylsilane [Silane, tetramethyl-]	75–76–3	10,000	g
Trichlorosilane [Silane, trichloro-]	10025–78–2	10,000	g
Trifluorochloroethylene [Ethene, chlorotrifluoro-]	79–38–9	10,000	f
Trimethylamine [Methanamine, N,N-dimethyl-]	75–50–3	10,000	f
Vinyl acetylene [1-Buten-3-yne]	689–97–4	10,000	f
Vinyl chloride [Ethene, chloro-]	75–01–4	10,000	a, f
Vinyl ethyl ether [Ethene, ethoxy-]	109–92–2	10,000	g
Vinyl fluoride [Ethene, fluoro-]	75–02–5	10,000	f
Vinylidene chloride [Ethene, 1,1-dichloro-]	75–35–4	10,000	g
Vinylidene fluoride [Ethene, 1,1-difluoro-]	75–38–7	10,000	f
Vinyl methyl ether [Ethene, methoxy-]	107–25–5	10,000	f

NOTE: Basis for Listing:

a Mandated for listing by Congress.

f Flammable gas.

g Volatile flammable liquid.

TABLE 4 TO § 68.130.—LIST OF REGULATED FLAMMABLE SUBSTANCES AND THRESHOLD QUANTITIES FOR ACCIDENTAL RELEASE PREVENTION

[CAS Number Order—63 Substances]

CAS No.	Chemical name	CAS No.	Threshold quantity (lbs)	Basis for listing
60–29–7	Ethyl ether [Ethane, 1,1'-oxybis-]	60–29–7	10,000	g
74–82–8	Methane	74–82–8	10,000	f
74–84–0	Ethane	74–84–0	10,000	f
74–85–1	Ethylene [Ethene]	74–85–1	10,000	f
74–86–2	Acetylene [Ethyne]	74–86–2	10,000	f
74–89–5	Methylamine [Methanamine]	74–89–5	10,000	f
74–98–6	Propane	74–98–6	10,000	f
74–99–7	Propyne [1-Propyne]	74–99–7	10,000	f
75–00–3	Ethyl chloride [Ethane, chloro-]	75–00–3	10,000	f
75–01–4	Vinyl chloride [Ethene, chloro-]	75–01–4	10,000	a, f
75–02–5	Vinyl fluoride [Ethene, fluoro-]	75–02–5	10,000	f
75–04–7	Ethylamine [Ethanamine]	75–04–7	10,000	f
75–07–0	Acetaldehyde	75–07–0	10,000	g
75–08–1	Ethyl mercaptan [Ethanethiol]	75–08–1	10,000	g
75–19–4	Cyclopropane	75–19–4	10,000	f
75–28–5	Isobutane [Propane, 2-methyl]	75–28–5	10,000	f
75–29–6	Isopropyl chloride [Propane, 2-chloro-]	75–29–6	10,000	g
75–31–0	Isopropylamine [2-Propanamine]	75–31–0	10,000	g
75–35–4	Vinylidene chloride [Ethene, 1,1-dichloro-]	75–35–4	10,000	g
75–37–6	Difluoroethane [Ethane, 1,1-difluoro-]	75–37–6	10,000	f
75–38–7	Vinylidene fluoride [Ethene, 1,1-difluoro-]	75–38–7	10,000	f
75–50–3	Trimethylamine [Methanamine, N, N-dimethyl-]	75–50–3	10,000	f
75–76–3	Tetramethylsilane [Silane, tetramethyl-]	75–76–3	10,000	g
78–78–4	Isopentane [Butane, 2-methyl-]	78–78–4	10,000	g
78–79–5	Isoprene [1,3,-Butadiene, 2-methyl-]	78–79–5	10,000	g
79–38–9	Trifluorochloroethylene [Ethene, chlorotrifluoro-]	79–38–9	10,000	f
106–97–8	Butane	106–97–8	10,000	f
106–98–9	1-Butene	106–98–9	10,000	f
106–99–0	1,3-Butadiene	106–99–0	10,000	f
107–00–6	Ethyl acetylene [1-Butyne]	107–00–6	10,000	f
107–01–7	2-Butene	107–01–7	10,000	f
107–25–5	Vinyl methyl ether [Ethene, methoxy-]	107–25–5	10,000	f
107–31–3	Methyl formate [Formic acid, methyl ester]	107–31–3	10,000	g
109–66–0	Pentane	109–66–0	10,000	g
109–67–1	1-Pentene	109–67–1	10,000	g
109–92–2	Vinyl ethyl ether [Ethene, ethoxy-]	109–92–2	10,000	g
109–95–5	Ethyl nitrite [Nitrous acid, ethyl ester]	109–95–5	10,000	f
115–07–1	Propylene [1-Propene]	115–07–1	10,000	f
115–10–6	Methyl ether [Methane, oxybis-]	115–10–6	10,000	f
115–11–7	2-Methylpropene [1-Propene, 2-methyl-]	115–11–7	10,000	f
116–14–3	Tetrafluoroethylene [Ethene, tetrafluoro-]	116–14–3	10,000	f
124–40–3	Dimethylamine [Methanamine, N-methyl-]	124–40–3	10,000	f
460–19–5	Cyanogen [Ethanedinitrile]	460–19–5	10,000	f
463–49–0	Propadiene [1,2-Propadiene]	463–49–0	10,000	f
463–58–1	Carbon oxysulfide [Carbon oxide sulfide (COS)]	463–58–1	10,000	f
463–82–1	2,2-Dimethylpropane [Propane, 2,2-dimethyl-]	463–82–1	10,000	f
504–60–9	1,3-Pentadiene	504–60–9	10,000	f
557–98–2	2-Chloropropylene [1-Propene, 2-chloro-]	557–98–2	10,000	g
563–45–1	3-Methyl-1-butene	563–45–1	10,000	f
563–46–2	2-Methyl-1-butene	563–46–2	10,000	g
590–18–1	2-Butene-cis	590–18–1	10,000	f
590–21–6	1-Chloropropylene [1-Propene, 1-chloro-]	590–21–6	10,000	g
598–73–2	Bromotrifluorethylene [Ethene, bromotrifluoro-]	598–73–2	10,000	f
624–64–6	2-Butene-trans [2-Butene, (E)]	624–64–6	10,000	f
627–20–3	2-Pentene, (Z)-	627–20–3	10,000	g
646–04–8	2-Pentene, (E)-	646–04–8	10,000	g
689–97–4	Vinyl acetylene [1-Buten-3-yne]	689–97–4	10,000	f
1333–74–0	Hydrogen	1333–74–0	10,000	f
4109–96–0	Dichlorosilane [Silane, dichloro-]	4109–96–0	10,000	f
7791–21–1	Chlorine monoxide [Chlorine oxide]	7791–21–1	10,000	f
7803–62–5	Silane	7803–62–5	10,000	f
10025–78–2	Trichlorosilane [Silane, trichloro-]	10025–78–2	10,000	g
25167–67–3	Butene	25167–67–3	10,000	f

Note: Basis for Listing: a Mandated for listing by Congress. f Flammable gas. g Volatile flammable liquid.

[59 FR 4493, Jan. 31, 1994. Redesignated at 61 FR 31717, June 20, 1996, as amended at 62 FR 45132, Aug. 25, 1997; 63 FR 645, Jan. 6, 1998]

Subpart G—Risk Management Plan

SOURCE: 61 FR 31726, June 20, 1996, unless otherwise noted.

§ 68.150 Submission.

(a) The owner or operator shall submit a single RMP that includes the information required by §§ 68.155 through 68.185 for all covered processes. The RMP shall be submitted in a method and format to a central point as specified by EPA prior to June 21, 1999.

(b) The owner or operator shall submit the first RMP no later than the latest of the following dates:

(1) June 21, 1999;

(2) Three years after the date on which a regulated substance is first listed under § 68.130; or

(3) The date on which a regulated substance is first present above a threshold quantity in a process.

(c) Subsequent submissions of RMPs shall be in accordance with § 68.190.

(d) Notwithstanding the provisions of §§ 68.155 to 68.190, the RMP shall exclude classified information. Subject to appropriate procedures to protect such information from public disclosure, classified data or information excluded from the RMP may be made available in a classified annex to the RMP for review by Federal and state representatives who have received the appropriate security clearances.

§ 68.155 Executive summary.

The owner or operator shall provide in the RMP an executive summary that includes a brief description of the following elements:

(a) The accidental release prevention and emergency response policies at the stationary source;

(b) The stationary source and regulated substances handled;

(c) The worst-case release scenario(s) and the alternative release scenario(s), including administrative controls and mitigation measures to limit the distances for each reported scenario;

(d) The general accidental release prevention program and chemical-specific prevention steps;

(e) The five-year accident history;

(f) The emergency response program; and

(g) Planned changes to improve safety.

§ 68.160 Registration.

(a) The owner or operator shall complete a single registration form and include it in the RMP. The form shall cover all regulated substances handled in covered processes.

(b) The registration shall include the following data:

(1) Stationary source name, street, city, county, state, zip code, latitude, and longitude;

(2) The stationary source Dun and Bradstreet number;

(3) Name and Dun and Bradstreet number of the corporate parent company;

(4) The name, telephone number, and mailing address of the owner or operator;

(5) The name and title of the person or position with overall responsibility for RMP elements and implementation;

(6) The name, title, telephone number, and 24-hour telephone number of the emergency contact;

(7) For each covered process, the name and CAS number of each regulated substance held above the threshold quantity in the process, the maximum quantity of each regulated substance or mixture in the process (in pounds) to two significant digits, the SIC code, and the Program level of the process;

(8) The stationary source EPA identifier;

(9) The number of full-time employees at the stationary source;

(10) Whether the stationary source is subject to 29 CFR 1910.119;

(11) Whether the stationary source is subject to 40 CFR part 355;

(12) Whether the stationary source has a CAA Title V operating permit; and

(13) The date of the last safety inspection of the stationary source by a Federal, state, or local government agency and the identity of the inspecting entity.

§ 68.165 Offsite consequence analysis.

(a) The owner or operator shall submit in the RMP information:

(1) One worst-case release scenario for each Program 1 process; and

(2) For Program 2 and 3 processes, one worst-case release scenario to represent all regulated toxic substances held above the threshold quantity and one worst-case release scenario to represent all regulated flammable substances held above the threshold quantity. If additional worst-case scenarios for toxics or flammables are required by §68.25(a)(2)(iii), the owner or operator shall submit the same information on the additional scenario(s). The owner or operator of Program 2 and 3 processes shall also submit information on one alternative release scenario for each regulated toxic substance held above the threshold quantity and one alternative release scenario to represent all regulated flammable substances held above the threshold quantity.

(b) The owner or operator shall submit the following data:

(1) Chemical name;

(2) Physical state (toxics only);

(3) Basis of results (give model name if used);

(4) Scenario (explosion, fire, toxic gas release, or liquid spill and vaporization);

(5) Quantity released in pounds;

(6) Release rate;

(7) Release duration;

(8) Wind speed and atmospheric stability class (toxics only);

(9) Topography (toxics only);

(10) Distance to endpoint;

(11) Public and environmental receptors within the distance;

(12) Passive mitigation considered; and

(13) Active mitigation considered (alternative releases only);

§ 68.168 Five-year accident history.

The owner or operator shall submit in the RMP the information provided in § 68.42(b) on each accident covered by §68.42(a).

§ 68.170 Prevention program/Program 2.

(a) For each Program 2 process, the owner or operator shall provide in the RMP the information indicated in paragraphs (b) through (k) of this section. If the same information applies to more than one covered process, the owner or operator may provide the information only once, but shall indicate to which processes the information applies.

(b) The SIC code for the process.

(c) The name(s) of the chemical(s) covered.

(d) The date of the most recent review or revision of the safety information and a list of Federal or state regulations or industry-specific design codes and standards used to demonstrate compliance with the safety information requirement.

(e) The date of completion of the most recent hazard review or update.

(1) The expected date of completion of any changes resulting from the hazard review;

(2) Major hazards identified;

(3) Process controls in use;

(4) Mitigation systems in use;

(5) Monitoring and detection systems in use; and

(6) Changes since the last hazard review.

(f) The date of the most recent review or revision of operating procedures.

(g) The date of the most recent review or revision of training programs;

(1) The type of training provided—classroom, classroom plus on the job, on the job; and

(2) The type of competency testing used.

(h) The date of the most recent review or revision of maintenance procedures and the date of the most recent equipment inspection or test and the equipment inspected or tested.

(i) The date of the most recent compliance audit and the expected date of completion of any changes resulting from the compliance audit.

(j) The date of the most recent incident investigation and the expected date of completion of any changes resulting from the investigation.

(k) The date of the most recent change that triggered a review or revision of safety information, the hazard review, operating or maintenance procedures, or training.

§ 68.175 Prevention program/Program 3.

(a) For each Program 3 process, the owner or operator shall provide the information indicated in paragraphs (b) through (p) of this section. If the same information applies to more than one covered process, the owner or operator may provide the information only once, but shall indicate to which processes the information applies.

(b) The SIC code for the process.

(c) The name(s) of the substance(s) covered.

(d) The date on which the safety information was last reviewed or revised.

(e) The date of completion of the most recent PHA or update and the technique used.

(1) The expected date of completion of any changes resulting from the PHA;

(2) Major hazards identified;

(3) Process controls in use;

(4) Mitigation systems in use;

(5) Monitoring and detection systems in use; and

(6) Changes since the last PHA.

(f) The date of the most recent review or revision of operating procedures.

(g) The date of the most recent review or revision of training programs;

(1) The type of training provided—classroom, classroom plus on the job, on the job; and

(2) The type of competency testing used.

(h) The date of the most recent review or revision of maintenance procedures and the date of the most recent equipment inspection or test and the equipment inspected or tested.

(i) The date of the most recent change that triggered management of change procedures and the date of the most recent review or revision of management of change procedures.

(j) The date of the most recent pre-startup review.

(k) The date of the most recent compliance audit and the expected date of completion of any changes resulting from the compliance audit;

(l) The date of the most recent incident investigation and the expected date of completion of any changes resulting from the investigation;

(m) The date of the most recent review or revision of employee participation plans;

(n) The date of the most recent review or revision of hot work permit procedures;

(o) The date of the most recent review or revision of contractor safety procedures; and

(p) The date of the most recent evaluation of contractor safety performance.

§ 68.180 Emergency response program.

(a) The owner or operator shall provide in the RMP the following information:

(1) Do you have a written emergency response plan?

(2) Does the plan include specific actions to be taken in response to an accidental releases of a regulated substance?

(3) Does the plan include procedures for informing the public and local agencies responsible for responding to accidental releases?

(4) Does the plan include information on emergency health care?

(5) The date of the most recent review or update of the emergency response plan;

(6) The date of the most recent emergency response training for employees.

(b) The owner or operator shall provide the name and telephone number of the local agency with which the plan is coordinated.

(c) The owner or operator shall list other Federal or state emergency plan requirements to which the stationary source is subject.

§ 68.185 Certification.

(a) For Program 1 processes, the owner or operator shall submit in the RMP the certification statement provided in § 68.12(b)(4).

(b) For all other covered processes, the owner or operator shall submit in the RMP a single certification that, to the best of the signer's knowledge, information, and belief formed after reasonable inquiry, the information submitted is true, accurate, and complete.

§ 68.190 Updates.

(a) The owner or operator shall review and update the RMP as specified in paragraph (b) of this section and submit it in a method and format to a central point specified by EPA prior to June 21, 1999.

(b) The owner or operator of a stationary source shall revise and update the RMP submitted under § 68.150 as follows:

(1) Within five years of its initial submission or most recent update required by paragraphs (b)(2) through (b)(7) of this section, whichever is later.

(2) No later than three years after a newly regulated substance is first listed by EPA;

(3) No later than the date on which a new regulated substance is first present in an already covered process above a threshold quantity;

(4) No later than the date on which a regulated substance is first present above a threshold quantity in a new process;

(5) Within six months of a change that requires a revised PHA or hazard review;

(6) Within six months of a change that requires a revised offsite consequence analysis as provided in § 68.36; and

(7) Within six months of a change that alters the Program level that applied to any covered process.

(c) If a stationary source is no longer subject to this part, the owner or operator shall submit a revised registration to EPA within six months indicating that the stationary source is no longer covered.

Subpart H—Other Requirements

SOURCE: 61 FR 31728, June 20, 1996, unless otherwise noted.

§ 68.200 Recordkeeping.

The owner or operator shall maintain records supporting the implementation of this part for five years unless otherwise provided in subpart D of this part.

§ 68.210 Availability of information to the public.

(a) The RMP required under subpart G of this part shall be available to the public under 42 U.S.C. 7414(c).

(b) The disclosure of classified information by the Department of Defense or other Federal agencies or contractors of such agencies shall be controlled by applicable laws, regulations, or executive orders concerning the release of classified information.

§ 68.215 Permit content and air permitting authority or designated agency requirements.

(a) These requirements apply to any stationary source subject to this part 68 and parts 70 or 71 of this chapter. The 40 CFR part 70 or part 71 permit for the stationary source shall contain:

(1) A statement listing this part as an applicable requirement;

(2) Conditions that require the source owner or operator to submit:

(i) A compliance schedule for meeting the requirements of this part by the date provided in § 68.10(a) or;

(ii) As part of the compliance certification submitted under 40 CFR 70.6(c)(5), a certification statement that the source is in compliance with all requirements of this part, including the registration and submission of the RMP.

(b) The owner or operator shall submit any additional relevant information requested by the air permitting authority or designated agency.

(c) For 40 CFR part 70 or part 71 permits issued prior to the deadline for registering and submitting the RMP and which do not contain permit conditions described in paragraph (a) of this section, the owner or operator or air permitting authority shall initiate permit revision or reopening according to the procedures of 40 CFR 70.7 or 71.7 to incorporate the terms and conditions consistent with paragraph (a) of this section.

(d) The state may delegate the authority to implement and enforce the requirements of paragraph (e) of this section to a state or local agency or agencies other than the air permitting authority. An up-to-date copy of any delegation instrument shall be maintained by the air permitting authority. The state may enter a written agreement with the Administrator under which EPA will implement and enforce the requirements of paragraph (e) of this section.

(e) The air permitting authority or the agency designated by delegation or agreement under paragraph (d) of this section shall, at a minimum:

(1) Verify that the source owner or operator has registered and submitted an RMP or a revised plan when required by this part;

(2) Verify that the source owner or operator has submitted a source certification or in its absence has submitted a compliance schedule consistent with paragraph (a)(2) of this section;

(3) For some or all of the sources subject to this section, use one or more mechanisms such as, but not limited to, a completeness check, source audits, record reviews, or facility inspections to ensure that permitted sources are in compliance with the requirements of this part; and

(4) Initiate enforcement action based on paragraphs (e)(1) and (e)(2) of this section as appropriate.

§ 68.220 Audits.

(a) In addition to inspections for the purpose of regulatory development and enforcement of the Act, the implementing agency shall periodically audit RMPs submitted under subpart G of this part to review the adequacy of such RMPs and require revisions of RMPs when necessary to ensure compliance with subpart G of this part.

(b) The implementing agency shall select stationary sources for audits based on any of the following criteria:

(1) Accident history of the stationary source;

(2) Accident history of other stationary sources in the same industry;

(3) Quantity of regulated substances present at the stationary source;

(4) Location of the stationary source and its proximity to the public and environmental receptors;

(5) The presence of specific regulated substances;

(6) The hazards identified in the RMP; and

(7) A plan providing for neutral, random oversight.

(c) Exemption from audits. A stationary source with a Star or Merit ranking under OSHA's voluntary protection program shall be exempt from audits under paragraph (b)(2) and (b)(7) of this section.

(d) The implementing agency shall have access to the stationary source, supporting documentation, and any area where an accidental release could occur.

(e) Based on the audit, the implementing agency may issue the owner or operator of a stationary source a written preliminary determination of necessary revisions to the stationary source's RMP to ensure that the RMP meets the criteria of subpart G of this part. The preliminary determination shall include an explanation for the basis for the revisions, reflecting industry standards and guidelines (such as AIChE/CCPS guidelines and ASME and API standards) to the extent that such standards and guidelines are applicable, and shall include a timetable for their implementation.

(f) *Written response to a preliminary determination.* (1) The owner or operator shall respond in writing to a preliminary determination made in accordance with paragraph (e) of this section. The response shall state the owner or operator will implement the revisions contained in the preliminary determination in accordance with the timetable included in the preliminary determination or shall state that the owner or operator rejects the revisions in whole or in part. For each rejected revision, the owner or operator shall explain the basis for rejecting such revision. Such explanation may include substitute revisions.

(2) The written response under paragraph (f)(1) of this section shall be received by the implementing agency within 90 days of the issue of the preliminary determination or a shorter period of time as the implementing agency specifies in the preliminary determination as necessary to protect public health and the environment. Prior to the written response being due and upon written request from the owner or operator, the implementing agency may provide in writing additional time for the response to be received.

(g) After providing the owner or operator an opportunity to respond under paragraph (f) of this section, the implementing agency may issue the owner or operator a written final determination of necessary revisions to the stationary source's RMP. The final determination may adopt or modify the revisions contained in the preliminary determination under paragraph (e) of this section or may adopt or modify the substitute revisions provided in the response under paragraph (f) of this section. A final determination that adopts a revision rejected by the owner or operator shall include an explanation of the basis for the revision. A final determination that fails to adopt a substitute revision provided under paragraph (f) of this section shall include an explanation of the basis for finding such substitute revision unreasonable.

(h) Thirty days after completion of the actions detailed in the implementation schedule set in the final determination under paragraph (g) of this section, the owner or operator shall be in violation of subpart G of this part and this section unless the owner or operator revises the RMP prepared under subpart G of this part as required by the final determination, and submits the revised RMP as required under § 68.150.

(i) The public shall have access to the preliminary determinations, responses, and final determinations under this section in a manner consistent with § 68.210.

(j) Nothing in this section shall preclude, limit, or interfere in any way with the authority of EPA or the state to exercise its enforcement, investigatory, and information gathering authorities concerning this part under the Act.

APPENDIX A TO PART 68—TABLE OF TOXIC ENDPOINTS

[As defined in §68.22 of this part]

CAS No.	Chemical name	Toxic endpoint (mg/L)
107–02–8	Acrolein [2-Propenal]	0.0011
107–13–1	Acrylonitrile [2-Propenenitrile]	0.076
814–68–6	Acrylyl chloride [2-Propenoyl chloride]	0.00090
107–18–6	Allyl alcohol [2-Propen-1-ol]	0.036
107–11–9	Allylamine [2-Propen-1-amine]	0.0032
7664–41–7	Ammonia (anhydrous)	0.14
7664–41–7	Ammonia (conc 20% or greater)	0.14
7784–34–1	Arsenous trichloride	0.010
7784–42–1	Arsine	0.0019
10294–34–5	Boron trichloride [Borane, trichloro-]	0.010
7637–07–2	Boron trifluoride [Borane, trifluoro-]	0.028
353–42–4	Boron trifluoride compound with methyl ether (1:1) [Boron, trifluoro[oxybis[methane]]-, T-4	0.023
7726–95–6	Bromine	0.0065
75–15–0	Carbon disulfide	0.16
7782–50–5	Chlorine	0.0087
10049–04–4	Chlorine dioxide [Chlorine oxide (ClO2)]	0.0028
67–66–3	Chloroform [Methane, trichloro-]	0.49
542–88–1	Chloromethyl ether [Methane, oxybis[chloro-]	0.00025
107–30–2	Chloromethyl methyl ether [Methane, chloromethoxy-]	0.0018
4170–30–3	Crotonaldehyde [2-Butenal]	0.029
123–73–9	Crotonaldehyde, (E)-, [2-Butenal, (E)-]	0.029
506–77–4	Cyanogen chloride	0.030
108–91–8	Cyclohexylamine [Cyclohexanamine]	0.16
19287–45–7	Diborane	0.0011
75–78–5	Dimethyldichlorosilane [Silane, dichlorodimethyl-]	0.026
57–14–7	1,1-Dimethylhydrazine [Hydrazine, 1,1-dimethyl-]	0.012
106–89–8	Epichlorohydrin [Oxirane, (chloromethyl)-]	0.076
107–15–3	Ethylenediamine [1,2-Ethanediamine]	0.49
151–56–4	Ethyleneimine [Aziridine]	0.018
75–21–8	Ethylene oxide [Oxirane]	0.090
7782–41–4	Fluorine	0.0039
50–00–0	Formaldehyde (solution)	0.012
110–00–9	Furan	0.0012
302–01–2	Hydrazine	0.011
7647–01–0	Hydrochloric acid (conc 37% or greater)	0.030
74–90–8	Hydrocyanic acid	0.011
7647–01–0	Hydrogen chloride (anhydrous) [Hydrochloric acid]	0.030
7664–39–3	Hydrogen fluoride/Hydrofluoric acid (conc 50% or greater) [Hydrofluoric acid]	0.016
7783–07–5	Hydrogen selenide	0.00066
7783–06–4	Hydrogen sulfide	0.042
13463–40–6	Iron, pentacarbonyl- [Iron carbonyl (Fe(CO)5), (TB-5–11)-]	0.00044
78–82–0	Isobutyronitrile [Propanenitrile, 2-methyl-]	0.14
108–23–6	Isopropyl chloroformate [Carbonochloridic acid, 1-methylethyl ester]	0.10
126–98–7	Methacrylonitrile [2-Propenenitrile, 2-methyl-]	0.0027

APPENDIX A TO PART 68—TABLE OF TOXIC ENDPOINTS—Continued

[As defined in §68.22 of this part]

CAS No.	Chemical name	Toxic end-point (mg/L)
74–87–3	Methyl chloride [Methane, chloro-]	0.82
79–22–1	Methyl chloroformate [Carbonochloridic acid, methylester]	0.0019
60–34–4	Methyl hydrazine [Hydrazine, methyl-]	0.0094
624–83–9	Methyl isocyanate [Methane, isocyanato-]	0.0012
74–93–1	Methyl mercaptan [Methanethiol]	0.049
556–64–9	Methyl thiocyanate [Thiocyanic acid, methyl ester]	0.085
75–79–6	Methyltrichlorosilane [Silane, trichloromethyl-]	0.018
13463–39–3	Nickel carbonyl	0.00067
7697–37–2	Nitric acid (conc 80% or greater)	0.026
10102–43–9	Nitric oxide [Nitrogen oxide (NO)]	0.031
8014–95–7	Oleum (Fuming Sulfuric acid) [Sulfuric acid, mixture with sulfur trioxide]	0.010
79–21–0	Peracetic acid [Ethaneperoxoic acid]	0.0045
594–42–3	Perchloromethylmercaptan [Methanesulfenyl chloride, trichloro-]	0.0076
75–44–5	Phosgene [Carbonic dichloride]	0.00081
7803–51–2	Phosphine	0.0035
10025–87–3	Phosphorus oxychloride [Phosphoryl chloride]	0.0030
7719–12–2	Phosphorus trichloride [Phosphorous trichloride]	0.028
110–89–4	Piperidine	0.022
107–12–0	Propionitrile [Propanenitrile]	0.0037
109–61–5	Propyl chloroformate [Carbonochloridic acid, propylester]	0.010
75–55–8	Propyleneimine [Aziridine, 2-methyl-]	0.12
75–56–9	Propylene oxide [Oxirane, methyl-]	0.59
7446–09–5	Sulfur dioxide (anhydrous)	0.0078
7783–60–0	Sulfur tetrafluoride [Sulfur fluoride (SF4), (T-4)-]	0.0092
7446–11–9	Sulfur trioxide	0.010
75–74–1	Tetramethyllead [Plumbane, tetramethyl-]	0.0040
509–14–8	Tetranitromethane [Methane, tetranitro-]	0.0040
7550–45–0	Titanium tetrachloride [Titanium chloride (TiCl4) (T-4)-]	0.020
584–84–9	Toluene 2,4-diisocyanate [Benzene, 2,4-diisocyanato-1-methyl-]	0.0070
91–08–7	Toluene 2,6-diisocyanate [Benzene, 1,3-diisocyanato-2-methyl-]	0.0070
26471–62–5	Toluene diisocyanate (unspecified isomer) [Benzene, 1,3-diisocyanatomethyl-]	0.0070
75–77–4	Trimethylchlorosilane [Silane, chlorotrimethyl-]	0.050
108–05–4	Vinyl acetate monomer [Acetic acid ethenyl ester]	0.26

[61 FR 31729, June 20, 1996, as amended at 62 FR 45132, Aug. 25, 1997]

Wednesday
January 6, 1999

Part IV

Environmental Protection Agency

40 CFR Part 68

Accidental Release Prevention Requirements; Risk Management Programs Under Clean Air Act Section 112(r)(7), Amendments; Final Rule

ENVIRONMENTAL PROTECTION AGENCY

40 CFR Part 68

[FRL–6214–9]

RIN 2050–AE46

Accidental Release Prevention Requirements; Risk Management Programs Under Clean Air Act Section 112(r)(7); Amendments

AGENCY: Environmental Protection Agency (EPA).

ACTION: Final rule.

SUMMARY: This action modifies the chemical accident prevention rule codified in 40 CFR Part 68. The chemical accident prevention rule requires owners and operators of stationary sources subject to the rule to submit a risk management plan (RMP) by June 21, 1999, to a central location specified by EPA. In this action, EPA is amending the rule to: add four mandatory and five optional RMP data elements, establish specific procedures for protecting confidential business information when submitting RMPs, adopt the government's use of a new industry classification system, and make technical corrections and clarifications to Part 68. However, as stated in the proposed rule for these amendments, this action does not address issues concerning public access to offsite consequence analysis data in the RMP.

DATES: The rule is effective February 5, 1999.

ADDRESSES: Supporting material used in developing the proposed rule and final rule is contained in Docket A–98–08. The docket is available for public inspection and copying between 8:00 a.m. and 5:30 p.m., Monday through Friday (except government holidays) at Room 1500, 401 M Street SW, Washington, DC 20460. A reasonable fee may be charged for copying.

FOR FURTHER INFORMATION CONTACT: Sicy Jacob or John Ferris, Chemical Emergency Preparedness and Prevention Office, Environmental Protection Agency (5104), 401 M Street SW, Washington, DC 20460, (202) 260–7249 or (202) 260–4043, respectively; or the Emergency Planning and Community Right-to-Know Hotline at 800–424–9346 (in the Washington, DC metropolitan area, (703) 412–9810). You may wish to visit the Chemical Emergency Preparedness and Prevention Office (CEPPO) Internet site, at www.epa.gov/ceppo.

SUPPLEMENTARY INFORMATION:

Regulated Entities

Entities potentially regulated by this action are those stationary sources that have more than a threshold quantity of a regulated substance in a process. Regulated categories and entities include:

Category	Examples of regulated entities
Chemical Manufacturers	Basic chemical manufacturing, petrochemicals, resins, agricultural chemicals, pharmaceuticals, paints, cleaning compounds.
Petroleum	Refineries.
Other Manufacturing	Paper, electronics, semiconductors, fabricated metals, industrial machinery, food processors.
Agriculture	Agricultural retailers.
Public Sources	Drinking water and waste water treatment systems.
Utilities	Electric utilities.
Other	Propane retailers and users, cold storage, warehousing, and wholesalers.
Federal Sources	Military and energy installations.

This table is not meant to be exhaustive, but rather provides a guide for readers to indicate those entities likely to be regulated by this action. The table lists entities EPA is aware of that could potentially be regulated by this action. Other entities not listed in the table could also be regulated. To determine whether a stationary source is regulated by this action, carefully examine the provisions associated with the list of substances and thresholds under § 68.130 and the applicability criteria under § 68.10. If you have questions regarding applicability of this action to a particular entity, consult the hotline or persons listed in the preceding FOR FURTHER INFORMATION CONTACT section.

Table of Contents

I. Introduction and Background

A. Statutory Authority

These amendments are being promulgated under sections 112(r) and 301(a)(1) of the Clean Air Act (CAA) as amended (42 U.S.C. 7412(r), 7601(a)(1)).

B. Background

The 1990 CAA Amendments added section 112(r) to provide for the prevention and mitigation of accidental chemical releases. Section 112(r) mandates that EPA promulgate a list of "regulated substances," with threshold quantities. Processes at stationary sources that contain a threshold quantity of a regulated substance are subject to accidental release prevention regulations promulgated under CAA section 112(r)(7). EPA promulgated the list of regulated substances on January 31, 1994 (59 FR 4478) (the "List Rule") and the accidental release prevention regulations creating the risk management program requirements on June 20, 1996 (61 FR 31668) (the "RMP Rule"). Together, these two rules are codified as 40 CFR Part 68. EPA amended the List Rule on August 25, 1997 (62 FR 45132), to change the listed concentration of hydrochloric acid. On January 6, 1998 (63 FR 640), EPA amended the List Rule to delist Division 1.1 explosives (classified by DOT), to clarify certain provisions related to regulated flammable substances and to clarify the transportation exemption.

Part 68 requires that sources with more than a threshold quantity of a regulated substance in a process develop and implement a risk management program that includes a five-year accident history, offsite consequence analyses, a prevention program, and an emergency response program. In Part 68, processes are divided into three categories (Programs 1 through 3). Processes that have no potential impact on the public in the case of accidental releases have minimal requirements (Program 1). Processes in Programs 2 and 3 have additional requirements based on the potential for offsite consequences associated with the worst-case accidental release and their accident history. Program 3 is also triggered if the processes are subject to OSHA's Process Safety Management (PSM) Standard. By June 21, 1999, sources must submit to a location designated by EPA, a risk management plan (RMP) that summarizes their implementation of the risk management program.

When EPA promulgated the risk management program regulations, it stated that it intended to work toward electronic submission of RMPs. The Accident Prevention Subcommittee of the CAA Advisory Committee convened an Electronic Submission Workgroup to examine technical and practical issues associated with creating a national electronic repository for RMPs. Based on workgroup recommendations, EPA is in the process of developing two systems, a user-friendly PC-based submission system (RMP*Submit) and a database of RMPs (RMP*Info).

The Electronic Submission Workgroup also recommended that EPA add some mandatory and optional data elements to the RMP and asked EPA to clarify how confidential business information (CBI) submitted in the RMP would be handled. Based on these recommendations and requests for clarifications, EPA proposed amendments to Part 68 on April 17, 1998 (63 FR 19216). These amendments proposed to replace the use of Standard Industrial Classification (SIC) codes with the North American Industry Classification System (NAICS) codes, add four mandatory data elements to the RMP, add five optional data elements to the RMP, establish specific requirements for submission of information claimed CBI, and make technical corrections and clarifications to the rule. EPA received 47 written comments on the proposed rule. Today's rule reflects EPA's consideration of all comments; major issues raised by commenters and EPA's responses are discussed in Section III of this preamble. A summary of all comments submitted and EPA's responses can be found in a document entitled, Accidental Release Prevention Requirements; Risk Management Programs Under Clean Air Act Section 112(r)(7); Amendments: Summary and Response to Comments, in the Docket (see ADDRESSES).

II. Summary of the Final Rule

NAICS Codes

On January 1, 1997, the U.S. Government, in cooperation with the governments of Canada and Mexico, adopted a new industry classification system, the North American Industry Classification System (NAICS), to replace the Standard Industrial Classification (SIC) codes (April 9, 1997, 62 FR 17288). The applicability of some Part 68 requirements (i.e., Program 3 prevention requirements) is determined, in part, by SIC codes, and Part 68 also requires the reporting of SIC codes in the RMP. Therefore, EPA is revising Part 68 to replace all references to "SIC code" with "NAICS code." In addition, EPA is replacing, as proposed, the nine SIC codes subject to Program 3 prevention program requirements with ten NAICS codes, as follows:

NAICS	Sector
32211	Pulp mills
32411	Petroleum refineries
32511	Petrochemical manufacturing
325181	Alkalies and chlorine
325188	All other inorganic chemical manufacturing
325192	Other cyclic crude and intermediate manufacturing
325199	All other basic organic chemical manufacturing
325211	Plastics and resins
325311	Nitrogen fertilizer
32532	Pesticide and other agricultural chemicals

NAICS codes are either five or six digits, depending on the degree to which the sector is subdivided.

RMP Data Elements

As proposed, EPA is adding four new data elements to the RMP: latitude/longitude method and description, CAA Title V permit number, percentage weight of a toxic substance in a liquid mixture, and NAICS code for each process that had an accidental release reported in the five-year accident history. EPA is also adding five optional data elements: local emergency planning committee (LEPC) name, source or parent company e-mail address, source homepage address, phone number at the source for public inquiries, and status under OSHA's Voluntary Protection Program (VPP).

Prevention Program Reporting

EPA is not revising Sections 68.170 and 68.175 as proposed. Prevention program reporting, therefore, will not be changed to require a prevention program for each portion of a process for which a Process Hazard Analysis (PHA) or hazard review was conducted. Instead, EPA plans to create functions within RMP*Submit to provide stationary sources with a flexible way of explaining the scope and content of each prevention program they implement at their facility.

Confidential Business Information

EPA is clarifying how confidential business information (CBI) submitted in the RMP will be handled. EPA has determined that the information required by certain RMP data elements does not meet the criteria for CBI and therefore may not be claimed as such. The Agency is also requiring submission of substantiation at the time a CBI claim is filed.

Finally, EPA is promulgating several of the technical corrections and clarifications, as proposed in the Federal Register, April 17, 1998 (63 FR 19216).

III. Discussion of Issues

EPA received 47 comments on the proposed rule. The commenters included chemical manufacturers, petroleum refineries, environmental groups, trade associations, a state agency, and members of the public. The major issues raised by commenters are addressed briefly below. The Agency's complete response to comments received on this rulemaking is available in the docket (see ADDRESSES). The document is titled Accidental Release Prevention Requirements; Risk Management Programs Under Clean Air Act Section 112(r)(7); Amendments: Summary and Response to Comments.

A. NAICS Codes

Two commenters asked that sources be given the option to use either SIC codes or NAICS codes, or both, in their initial RMP because the NAICS system is new and may not be familiar to sources. EPA disagrees with this suggestion. EPA intends to provide several outreach mechanisms to assist sources in identifying their new NAICS code. RMP*Submit will provide a "pick list" that will make it easier for sources to find the appropriate code. Also, selected NAICS codes are included in the General Guidance for Risk Management Programs (July 1998) and in the industry-specific guidance documents that EPA is developing. EPA will also utilize the Emergency Planning and Community Right-to-Know Hotline at 800–424–9346 (or 703–412–9810) and its web site at www.epa.gov/ceppo/, to assist sources in determining the source's NAICS codes. EPA also notes that the Internal Revenue Service is planning to require businesses to provide NAICS-based activity codes on their 1998 tax returns, so many sources will have become familiar with their NAICS codes by the June 1999 RMP deadline.

EPA believes it is necessary and appropriate to change from SIC codes to NAICS codes at this time. EPA recognizes that NAICS codes were developed for statistical purposes by the Office of Management and Budget (OMB). In the notice of April 9, 1997 (62 FR 17288) OMB stated that the "[u]se of NAICS for nonstatistical purposes (e.g., administrative, regulatory, or taxation) will be determined by the agency or agencies that have chosen to use the SIC for nonstatistical purposes." EPA has determined that NAICS is appropriate in this rule for several reasons. First, the reason the SIC codes were replaced by NAICS codes is because the SIC codes no longer accurately represent today's industries. The SIC codes will become more obsolete over time because OMB will no longer be supporting the SIC codes; therefore, no new or modified SIC codes will be developed to reflect future changes in industries. Second, as the SIC codes become obsolete, most users of SIC codes will likely change to NAICS codes over time, so future data sharing and consistency will be enhanced by use of NAICS codes in the RMP program. Third, through this rulemaking process, EPA has analyzed specific conversions of SIC codes to NAICS codes for the RMP program and was able to identify NAICS codes that were applicable to fulfilling the purposes of this rule. Finally, because the RMP reporting requirement is new, it is reasonable to begin the program with NAICS codes now rather than converting to them later.

Three commenters expressed support for the ten NAICS codes that EPA proposed to use in place of the nine SIC codes referenced in section 68.10(d)(1) of Part 68 and one commenter partially objected. Section 68.10(d)(1) provides that processes in the referenced codes are subject to Program 3 requirements (if not eligible for Program 1). One commenter objected to EPA's proposal to replace the SIC code for pulp and paper mills with only the NAICS code for pulp mills that do not also produce paper or paperboard. The commenter asked EPA to reexamine the accident history of paper and paperboard mills. As discussed in the preamble of the proposed rule, EPA reviewed the accident history data prior to proposing the new NAICS codes. Neither facilities that classify themselves as paper mills (NAICS Code 322121) nor paperboard mills (NAICS code 32213) met the accident history criteria that EPA used to select industrial sectors for Program 3.

EPA notes that a pulp process at a paper or a paperboard mill may still be subject to Program 3 as long as the process contains more than a threshold quantity of a regulated substance and is not eligible for Program 1. Section 68.10(d)(1) uses industrial codes to classify processes, not facilities as a whole. Since section 68.10(d)(1) will continue to list the code for pulp mills, pulpmaking processes will continue to be subject to Program 3. In addition, under section 68.10(d)(2), paper processes will be in Program 3 (unless eligible for Program 1) if they are subject to OSHA's Process Safety Management (PSM) standard. Most pulp and paper processes are, in fact, subject to this standard.

One commenter objected to assigning NAICS codes to a process rather than the source as a whole. EPA first notes that the requirement to assign a SIC code to a process was adopted in the original RMP rulemaking two years ago. Today's rule does not change that requirement except to substitute NAICS for SIC codes. In any event, EPA is today modifying Part 68 to clarify that sources provide the NAICS code that "most closely corresponds to the process." EPA believes that assigning an industry code to a process will help implementing agencies and the public understand what the covered process does; using the code makes it possible to provide this information without requiring a detailed explanation from the source. In addition, the primary NAICS code for a source as a whole may not reflect the activity of the covered process.

B. RMP Data Elements

EPA proposed to add, as optional RMP data elements: local emergency planning committee (LEPC), source (or parent company) E-mail address, source homepage address, phone number at the source for public inquiries, and OSHA Voluntary Protection Program (VPP) status. EPA also proposed to add, as mandatory data elements: method and description of latitude/longitude, Title V permit number, percent weight of a toxic substance in a liquid mixture, and NAICS code (only in the five-year accident history section).

Commenters generally supported the new optional data elements. One commenter requested that the optional elements be made mandatory. EPA disagrees with this comment. While the elements are useful, many sources covered by this rule will not have e-mail addresses or home pages. The RMP will provide both addresses and phone numbers so that the public will have methods to reach the source. EPA has learned that in some areas there are no functioning LEPCs, therefore, at this time, EPA will not add this as a mandatory data element. However, in most cases, the LEPC for an area can be determined by contacting the local government or the State Emergency Response Commission (SERC) for which the area is located. Therefore, reporting these data elements will remain optional at this time.

One commenter supported adding the listing of local emergency planning committee in the RMP data elements as an optional data element. The commenter stated that, although it is an optional data element, this listing will enhance the ability of local responders and emergency planners to adequately prepare and train for emergency events.

Of the data elements that were proposed to be mandatory, one commenter objected to the addition of latitude/longitude method and description. The commenter stated that it was not clear in the proposal why the method and description information is needed. EPA is seeking latitude/longitude method and description in accordance with its Locational Data Policy. Several EPA regulations require sources to provide their latitude and longitude, so that EPA can more readily locate facilities and communicate data between Agency offices. Sharing of data between EPA offices reduces duplication of information. Latitude/longitude method and description provides information needed by EPA offices, and other users of the data, to rectify discrepancies that may appear in the latitude and longitude information provided by the source under various EPA requirements. Documentation of the method by which the latitude and longitude are determined and a description of the location point referenced by the latitude and longitude (e.g., administration building) will permit data users to evaluate the accuracy of those coordinates, thus addressing EPA data sharing and integration objectives.

EPA believes this information will also facilitate EPA-State coordination of environmental programs, including the chemical accident prevention rule. The State/EPA Data Management Program is a successful multi-year initiative linking State environmental regulatory agencies and EPA in cooperative action. The Program's goals include improvements in data quality and data integration based on location identification. Therefore, as proposed, the latitude/longitude method and description will be added to the existing RMP data elements. RMP*Submit will provide a list of methods and descriptions from which sources may choose.

EPA also proposed to require that sources report the percentage weight (weight percent) of a toxic substance in a mixture in the offsite consequence analysis (OCA) and the accident history sections of the RMP. This information is necessary for users of RMP data to understand how worst case and alternative release scenarios have been modeled. EPA has decided to require reporting of the weight percent of toxic substance in a liquid mixture because this information is necessary to understand the volatilization rate, which determines the downwind dispersion distance of the substance. The volatilization rate is affected by the vapor pressure of the substance in the mixture. For example, a spill of 70 percent hydrofluoric acid (HF) will volatilize more quickly than a spill of the same quantity of HF in a 50 percent solution; consequently, over a 10-minute period, the 70 percent solution will travel further. Reviewers of the RMP data, including local emergency planning committees, need to know the weight percent to be able to evaluate the results reported in the offsite consequence analysis and the impacts reported in the accident history. Without knowing the weight percent of the substance in the mixture, users of the data may compare scenarios or incidents that appear to involve the same chemical in the same physical state, but in fact involve the same chemical held in a different physical state.

One commenter stated that for gas mixtures, percentage by volume (or volume percent) should be required to be reported rather than weight percent. In this final rule, EPA does not require reporting of the weight percent (or volume percent) of a regulated substance in a gas mixture. If a source handles regulated substances in a gaseous mixture (e.g., chlorine with hydrogen chloride), the quantity of a particular regulated substance in the mixture is what is reported in the RMP, since that is what would be released into the air. Its percentage weight in the mixture is irrelevant.

Another commenter objected to this data element, claiming that it could result in reverse engineering and create a competitive disadvantage. EPA does not believe that this requirement would create a competitive disadvantage, since similar information is available to the public under Emergency Planning and Community Right-to-Know Act (EPCRA) of 1986. Even so, if it were to have such an effect, sources can claim this element as CBI if it can meet the criteria for CBI claims in 40 CFR Part 2. Another commenter stated that the public would be concerned if the percentages did not add to 100, in the event that the source handles both regulated and non-regulated substances. EPA believes that because a source must model only one substance in a release scenario, the source need not report the percentages of the other substances in the mixture. Therefore, it is expected that the weight percent for mixtures would not always add up to 100, because the mixture could contain non-regulated substances.

A third commenter suggested that requiring sources to report percentage weight of a toxic substance in a liquid mixture would create confusion with the reporting of mixtures containing flammable regulated substances.

In the January 6, 1998 rule (63 FR 640), EPA clarified that flammable regulated substances in mixtures are only covered by the RMP rule if the entire mixture meets the National Fire Protection Association (NFPA) criteria of 4, thus the entire mixture becomes the regulated substance. As a result, the percentage of flammables in a mixture is not relevant under the rule and the requirement to report the percentage weight will only apply to toxic substances in a liquid mixture.

Finally, in the **Federal Register** notice of June 20, 1996 (61 FR 31688), EPA clarified the relationship between the risk management program and the air permit program under Title V of the CAA for sources subject to both requirements. Under section 502(b)(5)(A), permitting authorities must have the authority to assure compliance by all covered sources with each applicable CAA standard, regulation or requirement, including the regulations implementing section 112(r)(7). Requiring sources covered by Title V and section 112(r) to provide their Title V permit number will help Title V permitting authorities assure that each source is complying with the RMP rule.

In summary, with the exception of adding the phrase "that most closely corresponds to the process" in sections 68.42(b)(4), 68.160(b)(7), 68.170(b), and 68.175(b), EPA has decided to finalize the optional and mandatory data elements as they were proposed.

C. Prevention Program Reporting

The final RMP rule, issued June 20, 1996 (61 FR 31668), requires sources to report their prevention program for each "process." Because the applicable definition of "process" is broad, multiple production and storage units might be a single, complex "process." However, the Agency realizes that some elements of a source's prevention program for a process may not be applicable to every portion of the process. In such a situation, reporting prevention program information for the process as a whole could be misleading without an explanation of which prevention program element applies to which part of the process. In order to get more specific information on which prevention program practices apply to different production and storage units within a process, EPA proposed to revise the rule to require prevention program reporting for each part of the process for which a separate process hazard analysis (PHA) or hazard review was conducted. EPA further proposed deleting the second sentence from both sections 68.170(a) and 68.175(a), which presently states that, "[i]f the same information applies to more than one covered process, the owner or operator may provide the information only once, but shall indicate to which process the information applies."

A number of industry commenters objected to the proposed revisions as wrongly assuming that a one-to-one relationship exists between a prevention program and a PHA. The commenters asserted that EPA's proposed revision did not reflect how facilities conduct PHAs or implement prevention measures and would cause significant duplicate reporting, creating unnecessary extra work for facility personnel. One commenter explained that depending on a source's circumstances, it might conduct a PHA for each production line, including all of its different units, or it might conduct a PHA for each common element of its different production lines. Accordingly, the commenters claimed that EPA's proposal to require the owner/operator to submit separate prevention program information for every portion of a process covered by a PHA would result in multiple submissions of much of the same material, and would add no value to process safety or accidental release prevention. Commenters also opposed the deletion of the second sentence in sections 68.170(a) and 68.175(a). One commenter noted that many of the elements of the prevention program will not only be common to a process, but will be common to an entire stationary source. Thus commenters argued that EPA's proposals would result in redundant submittals and place an unjustified burden on the regulated community.

EPA acknowledges that PHAs do not necessarily determine the scope of prevention program measures. Moreover, EPA agrees that duplicative reporting should be reduced as much as possible. At the same time, EPA, implementing agencies, and other users of RMP data need to have information that is detailed enough to understand the hazards posed by, and the safety practices used for, particular parts of processes and equipment. EPA recognizes that some aspects of prevention programs are likely to be implemented facility-wide, rather than on a process or unit basis, whereas other aspects may apply to a particular process or only to particular units within a process. For example, most sources are likely to develop an employee participation plan and a system for hot work permits facility-wide, rather than on a process or unit basis. For sources having processes that include several units (e.g., multiple reactors or purification systems), the hazards, process controls, and mitigation systems may vary among the individual units. For example, one may have a deluge fire control system while another may have a runaway reaction quench system.

EPA has concluded that its proposed changes to prevention program reporting would not lead sources to prepare RMPs that accurately and efficiently communicate the hazards posed by different aspects of covered processes and the safety practices used to address those hazards. The Agency now believes that no rule changes are necessary to ensure that RMPs convey that information. The current rule already requires prevention program reporting, and the issue has been how to efficiently convey that information in sufficient detail. EPA believes that its electronic program for submitting RMPs can be designed to provide for sufficient specificity in prevention program reporting without requiring duplicative reporting. In particular, the Agency plans to create a comment/text field in RMP*Submit for specifying which parts of a prevention program apply to which portions of a particular process. For example, if a deluge system only applies to a certain part of the overall process, the source would indicate in the comment/text screen the portions of the process to which the deluge system applies.

To reduce the burden of reporting, EPA also plans to create a function in RMP*Submit which will allow a source to automatically copy prevention program data previously entered for one process to fill blank fields in another process's prevention program. The source could then edit any of the data elements that are different. For example, where the prevention programs for two processes are identical (e.g., two identical storage tanks that are considered separate processes), the source could copy the data entered for one to fill in the blank field for the other. If some of the data elements vary between the prevention programs, the source will be able to autofill and change only those items that vary among processes or units.

Although the autofill option will minimize the burden of reporting common data elements for those sources filing electronically, EPA has decided not to delete the sentence, in both sections 68.170(a) and 68.175(a), which states, "[i]f the same information applies to more than one covered process, the owner or operator may provide the information only once, but shall indicate to which processes the information applies", as proposed.

D. Confidential Business Information (CBI)

1. Background

A central element of the chemical accident prevention program as established by the Clean Air Act and implemented by Part 68 is providing state and local governments and the public with information about the risk of chemical accidents in their communities and what stationary sources are doing to prevent such accidents. As explained in the preamble to the final RMP rule (61 FR 31668, June 20, 1996), every covered stationary source is required to develop and implement a risk management program and provide information about that program in its RMP. Under CAA section 112(r)(7)(B)(iii), a source's RMP must be registered with EPA and also submitted to the Federal Chemical Safety and Hazard Investigation Board ("the Board"), the state in which the source is located, and any local entity responsible for emergency response or planning. That section also provides that RMPs "shall be available to the public under section 114(c)" of the CAA. Section 114(c) gives the public access to information obtained under the Clean Air Act except for information (other than emission data) that would divulge trade secrets.

As noted previously, in the final RMP rule EPA announced its plan to develop a centralized system for submitting electronic versions of RMPs that would reduce the paperwork burden on both industry and receiving agencies and provide ready public access to RMP data. Under the system, a covered source would submit its RMP on computer diskette, which would be entered into a central database that all interested parties could access electronically. The system would thus make it possible for a single RMP submission to reach all interested parties, including those identified in section 112(r)(7)(B)(iii).[1]

An important assumption underlying the Agency's central submission plan was that RMPs would rarely, if ever, contain confidential business information (CBI). Following publication of the final rule, concerns were raised that at least some of the information required to be reported in RMPs could be CBI in the case of particular sources. While the June 20, 1996 rule provided for protection of CBI under section 114(c) (see section 68.210(a)), EPA was asked to address how CBI would be protected in the context of the electronic programs being developed for RMP submission and public access.

In the April 17, 1998 proposal to revise the RMP rule, EPA made several proposals concerning protection of CBI. It first reviewed the information requirements for RMPs (sections 68.155–185) and proposed to find that certain required data elements would not entail divulging information that could meet the test for CBI set forth in the Agency's comprehensive CBI regulations at 40 CFR Part 2.[2] Information provided in response to those requirements could not be claimed CBI. EPA also requested comment on whether some information that might be claimed as CBI (e.g., worst-case release rate or duration) would be "emission data" and thus publicly available under section 114(c) even if CBI.

EPA administers a variety of statutes pertaining to the protection of the environment, each with its own data collection requirements and requirements for disclosure of information to the public. In the implementation of these statutes, the Agency collects emission, chemical, process, waste stream, financial, and other data from facilities in many, if not most, sectors of American business. Companies may consider some of this information vital to their competitive

[1] It is important to note that, as discussed in Section III. E of this preamble, this rule does not address issues concerning public access to offsite consequence analysis data in the RMP.

[2] Information is CBI if (1) the business has asserted a claim which has not expired, been waived, or been withdrawn; (2) the business has shown that it has taken and will continue to take reasonable steps to protect the information from disclosure; (3) the information is not and has not been reasonably obtainable by the public (other than governmental bodies) by use of legitimate means; (4) no statute requires disclosure of the information; and (5) disclosure of the information is likely to cause substantial harm to the business' competitive position. 40 CFR section 2.208.

position, and claim it as confidential business information (CBI).

In the course of implementing statutes, the Agency may have a need to communicate some or all of the information it collects to the public as the basis for a rulemaking, to its contractors, or in response to requests pursuant to the Freedom of Information Act (FOIA). Information found to be CBI is exempt from disclosure under FOIA. To manage both CBI claims and FOIA requests, EPA has promulgated in 40 CFR Part 2, Subpart B a set of procedures for reviewing CBI claims, releasing information found not to be CBI, and where authorized, disclosing CBI. Subpart B lists the criteria that information must meet in order to be considered CBI, as well as the special handling requirements the Agency must follow when disclosing CBI to authorized representatives.

For RMP requirements that might entail divulging CBI, EPA proposed that a source be required to substantiate a CBI claim to EPA at the time that it makes the claim. Under EPA's Part 2 regulations, a source claiming CBI generally is required to substantiate the claim only when EPA needs to make the information public as part of some proceeding (e.g., a rulemaking) or EPA receives a request from the public (e.g., under the Freedom of Information Act (FOIA)) for the information. In view of the public information function of RMPs and the interest already expressed by members of the public in them, EPA proposed "up-front substantiation" of CBI claims to ensure that information not meeting CBI criteria would be made available to the public as soon as possible. This approach of requiring up-front substantiation is the same as that used for trade secret claims filed under the Emergency Planning and Community Right-to-Know Act (EPCRA) of 1986.[3]

In addition, EPA proposed that any source claiming CBI submit two versions of its RMP: (1) a redacted ("sanitized"), electronic version, which would become part of RMP*Info, and (2) an unsanitized (unredacted) paper copy of the RMP (see proposed section 68.151(c)). The electronic database of RMPs would contain only the redacted version unless and until EPA ruled against all or part of the source's CBI claim, in keeping with the Part 2 procedures. In this way, the public would have access only to the non-CBI elements of sources' RMPs. EPA further stated that state and local agencies could receive the unredacted RMPs by requesting them from EPA under the Part 2 regulations. Those regulations authorize EPA to provide CBI to an agency having implementation responsibilities under the CAA if the agency either demonstrates that it has the authority under state or local law to compel such information directly from the source or that it will "provide adequate protection to the interests of affected businesses" (40 CFR 2.301(h)(3)).

The following sections of this preamble summarize and respond to the comments EPA received on the CBI-related aspects of its proposal. At the outset, however, EPA wants to emphasize that it does not anticipate many CBI claims being made in connection with RMPs. The Agency developed the RMP data elements with the issue of CBI in mind. It sought to define data elements that would provide basic information about a source's risk management program without requiring it to reveal CBI. To have done otherwise would have risked creating RMPs that were largely unavailable to the public. EPA continues to believe that the required RMP data elements will rarely require that a business divulge CBI. The Agency will carefully monitor the CBI claims made. If it appears that the number of claims being made is jeopardizing the public information objective of the chemical accident prevention program, EPA will consider ways of revising RMPs, including further rulemakings or revising the underlying program, to ensure that important health and safety information is available to the public.

2. RMP Data Elements Found Not CBI

Fifteen commenters representing environmental groups and members of the public opposed allowing some or all RMP data to be claimed as CBI in light of the public's interest in the information RMPs will provide. A number of commenters urged EPA not to allow the following RMP data elements (and supporting documents) to be claimed as CBI:

- Mitigation measures considered by the firm in its offsite consequence analysis,
- Major process hazards identified by the firm,
- Process controls in use,
- Mitigation systems in use,
- Monitoring and detection systems in use, and
- Changes since the last hazard review.

In addition, one commenter contended that even chemical identity and quantity should be ineligible for CBI protection, since the requirement to submit an RMP only applies to facilities using a few well-known, extremely hazardous chemicals, and the public's right to know should always outweigh a company's claim to CBI.

Along the same lines, a number of commenters urged EPA to develop a "corporate sunshine rule" that would allow confidentiality concerns to be overridden if the protected information is needed by the public and experts to understand and assess safety issues. Another commenter recommended that a business claiming a chemical's identity as CBI should be required to provide the generic name of the chemical and information about its adverse health effects so the public can determine the potential risks.

One commenter argued that some of the RMP data that EPA suggested could reveal CBI, (e.g., release rate), were not "emission data," because the worst case scenario data are theoretical estimates, and do not represent any real emissions, past or present.

Representatives of the chemical and petroleum industries disagreed with EPA's proposal to list the data elements that EPA believed could not reveal CBI in any case. These commenters asserted that EPA could not anticipate all the ways in which information required by a data element might reveal CBI, and accordingly urged the Agency to make

[3] Section 302 of EPCRA (codified in 40 CFR Part 355) requires any facility having more than a threshold planning quantity of an extremely hazardous substance (EHS) to notify its state emergency response commission (SERC) and local emergency planning committee (LEPC) that the facility is subject to emergency planning. The vast majority of toxic substances listed in 40 CFR Section 68.130 were taken from the EHS list. Section 303 of EPCRA requires LEPCs to prepare an emergency response plan for the community that is under their jurisdiction. Section 303 of EPCRA also requires that facilities subject to section 302 shall provide any information required by their LEPC necessary for developing and implementing the emergency plan. Section 304 of EPCRA requires an immediate notification of a release of an EHS or Hazardous Substances listed in 40 CFR Section 302.4 above a reportable quantity to state and local entities. Section 304 also requires a written follow-up which includes among other things, the chemical name, quantity released and any known or anticipated health risks associated with the release. Sections 311 and 312 of EPCRA (codified in 40 CFR Part 370) require facilities that are subject to OSHA Hazard Communication Standard (HCS), to provide information to its SERC, LEPC and local fire department. This information includes the hazards posed by its chemicals, and inventory information, including average daily amount, maximum quantity and general location. Section 313 of EPCRA (codified in 40 CFR Part 372) requires certain facilities that are in specific industries (including chemical manufacturers) and that manufacture, process, or otherwise use a toxic chemical above specified threshold amounts to report, among other things, the annual quantity of the toxic chemical entering each environmental medium. Most facilities covered by CAA 112(r) are covered by one or more of these sections of EPCRA. Section 322 of EPCRA (codified in Part 350) allows facilities to claim only the chemical identity as trade secret.

case-by-case determinations on CBI claims. They also contended that "emission data" under section 114(c) does not extend to data on possible, as opposed to actual, emissions, and thus that RMP information concerning potential accidental releases would not qualify as "emission data," which must be made available to the public.

As pointed out above, an important purpose of the chemical accident prevention program required by section 112(r) is to inform the public of the risk of accidents in their communities and the methods sources are employing to reduce such risks. EPA therefore believes that as much RMP data as possible should be available to the public as soon as possible. However, section 112(r)(7)(B)(iii) requires that RMPs be made "available to the public under section 114(c)," which provides for protection of trade secret information (other than emission data). Given the statute's direction to protect whatever trade secret information is contained in an RMP, EPA is not authorized to release such information even when the public's need for such information arguably outweighs a business' interest in its confidentiality. The Agency also cannot issue a "corporate sunshine rule" that conflicts with existing law requiring EPA (and other agencies) to protect trade secret information.

As explained above (and in more detail in the proposed rule), EPA examined each RMP data element to determine which would require information that might, depending on a business' circumstances, meet the CBI criteria set forth in EPA's regulations implementing section 114(c) and other information-related legal requirements. The point of this exercise was to both protect potential trade secret information and promote the public information purpose of RMPs by identifying which RMP information might reveal CBI in a particular case and by precluding CBI claims for information that could not reveal CBI in any case. EPA presented the results of its analysis and an explanation of why certain data elements could entail the reporting of CBI depending on a business' circumstances and why others could not. No commenter provided any specific examples or explanations that contradicted the Agency's rationale for its determinations of which data elements could or could not result in reporting of CBI.

However, EPA is deleting from the list of 40 CFR Part 68.151(b)(1) the reference to 40 CFR Part 68.160(b)(9), to allow for the possibility of the number of full-time employees at the stationary source to be claimed as CBI. Upon further review, EPA was unable to determine that providing the number of employees at the stationary source could never entail divulging information that could meet the test for CBI set forth in the Agency's comprehensive CBI regulations at 40 CFR Part 2. Therefore, EPA has removed this element from the list of data elements that can not be claimed CBI in Part 68. With this exception, EPA is promulgating the list of RMP data elements for which CBI claims are precluded, as proposed (Section 68.151(b)).

EPA's justifications for its specific CBI findings appear in an appendix to this preamble. A more detailed analysis of all RMP data elements and CBI determinations is available in the docket (see ADDRESSES). The Agency continues to find no reasonable basis for anticipating that the listed elements will in any case require a business to reveal CBI that is not "emission data." The information required by each of the listed data elements either fails to meet the criteria for CBI set forth in EPA's CBI regulations at Part 2 or meets the Part 2 definition of "emission data." In many cases, the information is available to the public through other reports filed with EPA, states, or local agencies (e.g., reports required by Emergency Planning and Community Right-to-Know Act (EPCRA) sections 312 and 313 provide general facility identification information and reports of most accidental releases are available through several Federal databases including EPA's Emergency Release Notification System and Accidental Release Information Program databases).

In order to preclude CBI claims for other data elements, the Agency would have to show that the information required by a data element either was "emission data" under section 114(c) or could not, under any circumstances, reveal CBI. As explained below, EPA does not believe such a showing can be made for any of the data elements not on the list. Therefore, CBI claims made for information required by data elements not on the list will be evaluated on a case-by-case basis according to the procedures contained in 40 CFR Part 2 (except that substantiation will have to accompany the claims, as discussed below).

The Agency agrees with the commenters who argued that information about potential accidental releases is not "emission data" under section 114(c). EPA's existing policy statement (see 56 FR 7042, Feb. 21, 1991) on what information may be considered "emission data" was developed to implement sections 110 and 114(a) of the CAA, which the Agency generally invokes when it seeks to gather technical data from a source about its actual emissions to the air. While the policy is not explicitly limited in its scope, EPA believes it would be inappropriate to apply it to RMP data elements concerning hypothetical, as opposed to actual, releases to the air. Under the definition of "emission data" contained in Part 2, information is "emission data" if it is (1) "necessary to determine the identity, amount, frequency, concentration, or other characteristics * * * of any emission *which has been emitted* by the source," (2) "necessary to determine the identity, amount, frequency, concentration, or other characteristics * * * of the emissions which, under an applicable standard or limitation, the source was *authorized to emit*;" or (3) general facility identification information regarding the source which distinguishes it from other sources (40 CFR section 2.301(a)(2)(i) (emphasis added)). Under these criteria, EPA has concluded that only the RMP data elements relating to source-level registration information (sections 68.160(b)(1)–(6), (8)–(13)) and the five-year accident history (section 68.168) are "emission data." Of the RMP data elements, only the five-year accident history involves actual, past emissions to the environment; the other data elements would not, therefore, qualify as "emission data" under the first prong of the Part 2 definition. Moreover, the data elements relating to a source's offsite consequence analysis, prevention program and emergency response program do not attempt to identify or otherwise reflect "authorized" emissions; the data elements instead reflect the source's *potential* for accidental releases. Accordingly, these data elements would not be "emission data" under the second prong of the definition. As for the third prong, some of the source-level data are "emission data" because they help identify a source. Most other RMP data elements are reported on a process level and are not generally used to distinguish one source from another.

The Agency believes it is unable to show that the remaining data elements could not, under any circumstances, reveal CBI. EPA continues to believe that it is theoretically possible for the remaining data elements (the elements not listed in section 68.151(b)) to reveal CBI either directly or through reverse engineering, depending on the circumstances of a particular case. At the same time, EPA believes that, in practice, the remaining data elements will rarely reveal CBI. The purpose of the data in the RMP is for a source to articulate its hazards, and the steps it takes to prevent accidental releases. In general, the kinds of information specifying the source's hazards and risk management program are not likely to be competitively sensitive.

In particular, covered processes at the vast majority of stationary sources subject to the RMP rule are too common and well-known to support a CBI claim for information related to such processes. For example, covered public drinking water and wastewater treatment plants generally use common regulated substances in standard processes (i.e., chlorine used for disinfection). Also, covered processes at many sources involve the storage of regulated substances that the sources sell (e.g., propane, ammonia), so the processes are already public knowledge. Other covered processes involve the use of well-known combinations of regulated substances such as refrigerants. RMP information regarding these types of processes should not include CBI.

Even in the case of unusual or unique processes, it is generally unlikely that RMP information could be used to reveal CBI through reverse engineering. To begin with, required RMP information is general enough that it is unlikely to provide a basis for reverse engineering a process. For example, a source must report in its RMP whether overpressurization is a hazard and whether relief valves are used to control pressure, but it is not required to report information on actual pressures used, flow rates, chemical composition, or the configuration of equipment. Moreover, while RMP information may provide some data that could be used in an attempt to discover CBI information through reverse engineering, it typically will not provide enough data for such an attempt to succeed, because the source is not required to provide a detailed description of the chemistry or production volume of the process. Businesses claiming CBI based on the threat of reverse engineering will be required to show how reverse engineering could in fact succeed with the information that the RMP would otherwise make public, together with other publicly available information. A business unable to do so will have its claim denied.

While EPA is requiring that a source claiming a chemical's identity as CBI provide the generic category or class name of the chemical, the RMP does not require sources to provide information about the adverse health effects of the chemical. Chemicals were included in the section 112(r) program because they are acutely toxic or flammable; health effects related to chronic exposure were not considered because they are addressed by other rules (see List Rule at 59 FR 4481). EPA believes that generic names are sufficient to indicate the general health concerns from short-term exposures. Should a member of the public desire more information, EPA encourages the use of EPCRA section 322(h), which provides a means for the public to obtain information about the adverse health effects of a chemical covered by that statute, where the chemical's identity has been claimed a trade secret. The public will find this provision of EPCRA useful because most sources subject to the RMP rule are also subject to EPCRA.

3. Up-front Substantiation of CBI Claims

One commenter supported the proposal to require CBI claims to be substantiated at the time they are made. Another commenter stated that there is no compelling need to require up-front substantiation. The commenter stated that up-front substantiation would place a sizable burden on both industry and EPA and would be in direct conflict with the Paperwork Reduction Act. The commenter claimed that, with the exception of EPCRA, where a submitter is allowed to claim only one data element—chemical identity—as CBI, it is EPA's standard procedure not to require submitters to provide written substantiation unless a record has been requested. Further, the commenter stated that the Agency has not shown any reason for departing from that procedure in this rule.

EPA believes that requiring up-front substantiation of CBI claims made for RMP data has ample precedent, is fully consistent with the Agency's CBI regulations and the Paperwork Reduction Act, and is critical to achieving the public information purposes of the accident prevention program. EPCRA is not the only example of an up-front substantiation requirement. The Agency has also required up-front substantiation in several other regulatory contexts, including those where, like here, providing the public with health and safety information is an important objective [see e.g., 40 CFR section 725.94, 40 CFR section 710.38, and 40 CFR section 720.85 (regulations promulgated under Toxic Substances Control Act)].

Even under its general CBI regulations, the Agency need not wait for a request to release data to require businesses to substantiate their CBI claims. When EPA expects to get a request to release data claimed confidential, the Agency is to initiate "at the earliest practicable time" the regulations" procedures for making CBI determinations (40 CFR section 2.204(a)(3)). Those procedures include calling on affected businesses to substantiate their claims (see 40 CFR section 2.204(e)). Since state and local agencies, environmental groups, academics and others have already indicated their interest in obtaining complete RMP data (see comments received on this rulemaking, available in the DOCKET), EPA fully expects to get requests for RMP data claimed CBI. Consequently, even if EPA did not establish an up-front substantiation requirement in this rule, under the Agency's general CBI regulations it could require businesses claiming CBI for RMP data to substantiate their claims without first receiving a request to release the data. Establishing an up-front requirement in this rule will simply allow EPA to obtain substantiation of CBI claims without having to request it in every instance.

Requiring up-front substantiation for RMP CBI claims is consistent with the Paperwork Reduction Act. Any burden posed by this requirement has already been evaluated as part of the Information Collection Request (ICR) associated with this rulemaking. EPA disagrees that up-front substantiation will impose a substantial or undue burden. As noted above, under EPA's current CBI regulations, a source claiming CBI could and probably would be required to provide substantiation for its claim, in view of the public interest in RMP information. A requirement to submit substantiation with the claim should thus make little difference to the source. Moreover, a source presumably does not make any claim of CBI lightly. Before filing a CBI claim, the source must first determine whether the claim meets the criteria specified in 40 CFR section 2.208. Up-front substantiation only requires that the source document that determination at the time it files its claim. Since it would be sensible for a source to document the basis of its CBI claim for its own purposes (e.g., in the case of a request for substantiation), EPA expects that many sources already prepare documentation for their CBI claims by the time they file them. Also, submitting substantiation at the time of claim reduces any additional burden later, such as reviewing the Agency's request, retrieving the relevant information, etc. Therefore, providing documentation at the time of filing should impose no additional burden.

In view of the public information function of RMPs, EPA believes that up-front substantiation is clearly warranted for CBI claims made for RMP data. Up-front substantiation will ensure that sources filing claims have carefully considered whether the data they seek to protect in fact meets the criteria for protection. Given the public interest already expressed in RMP data, EPA expects that CBI claims for RMP data will have to be substantiated at some point. Up-front substantiation will save EPA and the public time and resources that would otherwise be required to respond to each CBI claim with a request for substantiation. EPA is therefore promulgating the up-front substantiation requirement as proposed.

4. State and Local Agency Access to Unredacted RMPs

One commenter objected to EPA's statement in the proposal that it would provide unredacted (unsanitized) versions of the RMPs to a state and local agency only upon meeting the criteria required by the EPA's CBI rules at 40 CFR Part 2.[4] The commenter, an association of fire fighters, argued that the Agency's position was inconsistent with CAA section 112(r)(7)(B)(iii), which provides that RMPs "shall . . . be submitted to the Chemical Safety and Hazard Investigation Board [a federal agency], to the State in which the stationary source is located, and to any local agency or entity having responsibility for planning for or responding to accidental releases which may occur at such source . . . ." The commenter claimed that this provision entitles the specified entities, including local fire departments, to receive unredacted RMPs without having to make the showings required by EPA's CBI regulations.

EPA is not resolving this issue today. The Agency has reviewed the relevant statutory text and legislative history, as well as analogous provisions of EPCRA, and believes that arguments can be made on both sides of this issue. While section 112(r)(7)(B)(iii) calls for RMPs to be submitted to states, local entities and the Board, it is not clear that Congress intended CBI contained in RMPs to be provided to those entities without ensuring appropriate protection of CBI. At stake in resolving this issue are two important interests—local responders' interest in unrestricted access to information that may be critical to their safety and effectiveness in responding to emergencies and businesses' interest in protecting sensitive information from their competitors. Before making a final decision on this issue, EPA believes it would benefit from further public input. Because EPA stated that it would not provide unredacted RMPs to states and local agencies, those interested in protecting CBI may not have considered it necessary to lay out the legal and policy arguments supporting their views. State and local agencies, many of which in the past have expressed concern about the potential administrative burden of receiving RMPs directly from sources, also did not comment on the issue. EPA has therefore decided to accept additional comments on this issue alone. (Additional comments on any other issues addressed in this rulemaking will not be considered or addressed, since the Agency is taking final action on them here.) Comments should be mailed to the persons listed in the preceding FOR FURTHER INFORMATION CONTACT section. In the meantime, unredacted RMPs will be available to states, local agencies and the Board under the terms of the Agency's existing CBI regulations at 40 CFR section 2.301(h)(3) (for state and local agencies) and 40 CFR section 2.209(c) (for the Board).

Section 112(r)(7)(B)(iii) states in relevant part:

> [RMPs] shall also be submitted to the Chemical Safety and Hazard Investigation Board, to the State in which the stationary source is located, and to any local agency or entity having responsibility for planning for or responding to accidental releases which may occur at such source, and shall be available to the public under section 114(c) of [the Act].

Section 114(c) provides for the public availability of any information obtained by EPA under the Clean Air Act, except for information (other than emissions data) that would divulge trade secrets.

From a public policy perspective, there are some obvious advantages to reading section 112(r)(7)(B)(iii) in the way the commenter suggests. Local fire departments and other local responders are typically the first to arrive at the scene of chemical accidents in their jurisdictions. RMP information that first responders could find helpful include chemical identity, chemical quantity, and potential source of an accident. Under EPA's regulations, however, any or all of this information could be claimed CBI. In addition, state and local authorities are often in the best position to assess the adequacy of a source's risk management program and to initiate a dialogue with the facility should its RMP indicate a need for improvement. However, state and local authorities' ability to provide this contribution to community safety would be impeded to the extent a source claimed key information as CBI. While states and local agencies may obtain information claimed CBI under EPA's CBI regulations (assuming they can make the requisite showing), the time required to obtain the necessary authority or findings from state or local and EPA officials could be substantial.

At the same time, there are also public policy reasons for ensuring protection of CBI contained in RMPs. Congress has in many statutes, including the CAA and EPCRA, provided for the protection of trade secrets to safeguard the competitive position of private businesses. Businesses' ability to maintain the confidentiality of trade secrets helps ensure competition in the U.S. economy and U.S. businesses' competitive position in the world economy. Protection of trade secrets also encourages innovation, which is an important contributor to economic growth.

A reading of section 112(r)(7)(B)(iii) that demands submission of unredacted RMPs to states, local entities, and the Board may lead to widespread public access to information claimed CBI. For purposes of section 112(r)(7)(B)(iii), "any local agency or entity having responsibility for planning for or responding to accidental releases" includes local emergency planning committees (LEPCs) established under EPCRA. Section 301(c) of EPCRA provides that LEPCs must include representatives from both the public and private sectors, including the media and facilities subject to EPCRA requirements. Submission of an unredacted RMP to an LEPC would thus entail release of CBI to some members of the public and potentially even competitors.[5] More generally, local agencies may not be subject to any legal requirement to protect CBI and may lack the knowledge and resources to address CBI claims. Arguably, it would be

[4] Section 2.301(h)(3) provides that a State or local government may obtain CBI from EPA under two circumstances: (1) it provides EPA a written opinion from its chief legal officer or counsel stating that the State or local agency has the authority under applicable State or local law to compel the business to disclose the information directly; or (2) the businesses whose information is disclosed are informed and the State or local government has shown to a EPA legal office's satisfaction that its disclosure of the information will be governed by State or local law and by "procedures which will provide adequate protection to the interests of affected businesses."

[5] EPA does not believe that submission of an RMP containing CBI to the statutorily specified entities would defeat a source's ability to claim information as CBI for purposes of section 114(c) and EPA's CBI regulations. Under those regulations, information that has been released to the public cannot be claimed CBI. Release of a RMP containing CBI to the entities specified by section 112(r)(7)(B)(iii), including LEPCs, would not constitute such a release. EPCRA similarly provides that disclosure of trade secret information to an LEPC does not prevent a facility from claiming the information confidential (see EPCRA section 322(b)(1)).

anomalous for Congress to require EPA to protect trade secrets contained in RMPs against release to the public only to risk divulging the same information by requiring submission of unredacted RMPs to a broad range of entities that may not have the need or capacity to protect CBI themselves. It would also appear inconsistent with the approach Congress took to protecting trade secrets in EPCRA, where Congress did not provide for release of trade secret chemical identity information to local agencies.

Relatedly, many state and local agencies objected to EPA's original proposal in the RMP proposed rulemaking (58 FR 54190, October 20, 1993) that sources submit RMPs directly to States, local agencies, and the Board, as well as EPA. They noted that managing the information contained in RMPs would be difficult without a significant expenditure of typically scarce resources. Many states and local agencies thus supported EPA's final decision to develop an electronic submission and distribution system that would allow covered sources to submit their RMPs to EPA, which would make them available to states, local agencies, and the Board, as well as the general public. If the statute is read to require submission of RMP information to state and local agencies, and the Board, to the extent it is claimed as CBI, the resource concerns raised by State and local agencies commenters likely would be raised to that extent again.

EPA also questions the extent to which states, local entities and the Board would be disadvantaged if they did not receive unredacted RMPs without making the showings required by EPA's CBI regulations. As noted earlier, EPA expects that relatively little RMP information will be CBI. RMP data will only rarely contain CBI, and the up-front substantiation will minimize the number of CBI claims it receives by ensuring that sources carefully examine the basis for any claims before submitting them. Consequently, the Agency believes that a state or local agency will rarely confront a redacted RMP.

Moreover, EPCRA provides state and local entities, including fire departments, with access to much of the pertinent data already. EPA's regulations under EPCRA cover a universe of sources and chemicals that includes most, if not all, the sources and substances covered by the RMP rule. The EPCRA regulations require reporting of some of the same information required by the RMP rule, including chemical identity. EPCRA withholds from public release only chemical identities that are trade secrets and the location of specific chemicals where a facility so requests. In practice, relatively few facilities have requested trade secret protection for a chemical's identity.

Additionally, EPCRA section 312(f) empowers local fire departments to conduct on-site inspections at facilities subject to EPCRA section 312(a) and obtain information on chemical location. Most facilities subject to EPCRA section 312(a) are also subject to the RMP rule. On-site inspections could also provide information on hazards and mitigation measures. In addition, EPCRA section 303(d)(3) authorizes LEPCs, which include representatives of fire departments, to request from facilities covered by EPCRA section 302(b) such information as may be necessary to prepare an emergency response plan and to include such information in the plan as appropriate. Some sources subject to the RMP rule are also covered by EPCRA section 302(b).

In light of the points made above, EPA questions whether section 112(r)(7)(B)(iii) should be interpreted to require submission of unredacted RMPs containing CBI to the statutorily specified entities without provision being made for protecting CBI. EPA invites the public to provide any additional comment or information relevant to interpreting the submission requirement of section 112(r)(7)(B)(iii).

5. Other CBI Issues

Two commenters disagreed with EPA's statement that a source cannot make a CBI claim for information available to the public under EPCRA or another statute. They claimed that a request for information under EPCRA cannot supersede the CBI provisions applicable to data collected under the authorities of the CAA or Toxic Substances Control Act or any other regulatory program.

EPA does not agree with this comment. Claims of CBI may not be upheld if the information is properly obtainable or made public under other statutes or authorities. For example, chemical quantity on site is available to the public under EPCRA Tier II reporting. In addition, under EPCRA section 303(d)(3), LEPCs have the authority to request any information they need to develop and implement community emergency response plans. If information obtained through such a request is included in the community plan, it will become available to the public under EPCRA section 324. Information obtainable or made public under EPCRA would not be eligible for CBI protection under 40 CFR section 2.208, which specifically excludes from CBI protection information already available to the public. Filing a CBI claim under the CAA or another statute does not protect information if it is legitimately requested and made public under other federal, state, or local law. Information obtainable or made public (through proper means) under existing statutes cannot be CBI under EPA's CBI regulations.

6. Actions Taken

In summary, the Agency is adding two sections (68.151 and 68.152) to Part 68. Section 68.151 sets forth the procedures for a source to follow when asserting a CBI claim and lists data elements that can not be claimed as CBI. This section also requires sources filing CBI claims to provide the information claimed confidential, in a format to be specified by EPA, instead of the unsanitized paper copy of the RMP as discussed in the proposal. Section 68.152 sets forth the procedures for substantiating CBI claims. Sources claiming CBI are required to submit their substantiation of their claims at the same time they submit their RMPs.

E. Other Issues

Two commenters asked why EPA had proposed to drop the phrase "if used" in section 68.165(b)(3) where the rule asks for the basis of the offsite consequence analysis results. EPA has decided to retain the language, since sources will have a choice of using either EPA's RMP guidance documents or a model. Where a model is used, the source will have to provide the name of the model. These commenters also asked why EPA proposed to drop (alternative releases only) from section 68.165(b)(13). EPA has also decided to retain the parenthetical language.

One commenter stated that EPA should allow sources to submit RMPs either electronically or in hard copy. The commenter stated that not allowing hard copy submissions will be burdensome on many sources who have never filed an electronic report to the government before. As stated in the April proposal, EPA is allowing sources to submit RMPs on paper. Paper submitters are asked to fill out a simple paper form to tell EPA why they are unable to file electronically.

Two commenters objected to placing offsite consequence analysis (OCA) data, particularly worst-case release scenarios, on the Internet, for security reasons. Issues related to public access to OCA data are beyond the scope of this rulemaking, as this action is limited to the issues discussed above. It does not include decisions regarding how the public will access the OCA data elements of the RMPs. Statements in the preamble about EPA providing public access to RMP data are not intended to address which portions of the RMP data will be electronically available.

A number of commenters were concerned about a statement EPA made in the preamble to the proposed rule regarding the definition of "process", and stated that EPA's interpretation of "process" is not consistent with the interpretation the Occupational Safety and Health Administration (OSHA) uses in its process safety management (PSM) standard (29 CFR 1910.119). In this rulemaking, EPA did not propose any changes to the definition of process nor is it adopting any changes to the definition. As EPA stated in the preamble to the final RMP rule, it will interpret "process" consistently with OSHA's interpretation of that term (29 CFR 1910.119). Therefore, if a source is subject to the PSM rule, the limits of its process(es) for purposes of OSHA PSM will be the limits of its process(es) for purposes of RMP (except in cases involving atmospheric storage tanks containing flammable regulated substances, which are exempt from PSM but not RMP). If a source is not covered by OSHA PSM and is complicated from an engineering perspective, it should consider contacting its implementing agency for advice on determining process boundaries. EPA and OSHA are coordinating the agencies' approach to common issues, such as the interpretation of "process".

F. Technical Corrections

When Part 68 was promulgated, the text of section 68.79(a), was drawn from the OSHA PSM standard, but it was not revised to reflect the different structure of EPA's rule. The OSHA PSM standard is contained in a single section; EPA's Program 3 prevention program is contained in a subpart. Rather than referencing "this section," the paragraph should have referenced the "subpart." Therefore, as proposed, EPA is changing "section" to "subpart" in section 68.79(a).

Under section 68.180(b), EPA intended that all covered sources report the name and telephone number of the agency with which they coordinate emergency response activities, even if the source is not required to have an emergency response plan. However, the rule refers only to coordinating the emergency plan. In this action, EPA is revising this section to refer to the local agency with which emergency response activities and the emergency response plan is coordinated.

IV. Section-by-Section Discussion of the Final Rule

In Section 68.3, Definitions, the definition of SIC is removed and replaced by the definition of NAICS.

Section 68.10, Applicability, is revised to replace the SIC codes with NAICS codes, as discussed above.

Section 68.42, Five-Year Accident History, is revised to require the percentage concentration by weight of regulated toxic substances released in a liquid mixture and the five- or six-digit NAICS code that most closely corresponds to the process that had the release. The phrase "five- or six-digit" has been added before the NAICS code to clarify the level of detail required for NAICS code reporting.

Section 68.79, Compliance Audits, the word "section" in paragraph (a) is replaced by "subpart."

Section 68.150, Submission, is revised by adding a paragraph to state that procedures for asserting CBI claims and determining the sufficiency of such claims are provided in new Sections 68.151 and 68.152.

Section 68.151 is added to set forth the procedures to assert a CBI claim and list data elements that may not be claimed as CBI, as discussed above.

Section 68.152 is added to set forth procedures for substantiating CBI claims, as proposed.

Section 68.160, Registration, is revised by adding the requirements to report the method and description of latitude and longitude, replacing SIC codes with five- or six-digit NAICS codes, and adding the requirement to report Title V permit number, when applicable. This section is also revised to include optional data elements. The phrase "five- or six-digit" has been added before NAICS code to clarify the level of detail required for NAICS code reporting.

Section 68.165, Offsite Consequence Analysis, is revised by adding the requirement that the percentage weight of a regulated toxic substance in a liquid mixture be reported.

Section 68.170, Prevention Program/Program 2, is revised to replace SIC codes with five- or six-digit NAICS codes, as is Section 68.175.

Section 68.180, Emergency Response Program, is revised to clarify that paragraph (b) covers both the coordination of response activities and plans, as proposed.

V. Judicial Review

The proposed rule amending the accidental release prevention requirements; under section 112(r)(7) was proposed in the Federal Register on April 17, 1998. This Federal Register action announces EPA's final decision on the amendments. Under section 307(b)(1) of the CAA, judicial review of this action is available only by filing a petition for review in the U.S. Court of Appeals for the District of Columbia Circuit on or before March 8, 1999. Under section 307(b)(2) of the CAA, the requirements that are the subject of today's action may not be challenged later in civil or criminal proceedings brought by EPA to enforce these requirements.

VI. Administrative Requirements

A. Docket

The docket is an organized and complete file of all the information considered by the EPA in the development of this rulemaking. The docket is a dynamic file, because it allows members of the public and industries involved to readily identify and locate documents so that they can effectively participate in the rulemaking process. Along with the proposed and promulgated rules and their preambles, the contents of the docket serve as the record in the case of judicial review. (See section 307(d)(7)(A) of the CAA.)

The official record for this rulemaking, as well as the public version, has been established for this rulemaking under Docket No. A–98–08 (including comments and data submitted electronically). A public version of this record, including printed, paper versions of electronic comments, which does not include any information claimed as CBI, is available for inspection from 8:00 a.m. to 5:30 p.m., Monday through Friday, excluding legal holidays. The official rulemaking record is located at the address in ADDRESSES at the beginning of this document.

B. Executive Order 12866

Under Executive Order (E.O.) 12866, [58 FR 51,735 (October 4, 1993)], the Agency must determine whether the regulatory action is "significant", and therefore subject to OMB review and the requirements of the E.O. The Order defines "significant regulatory action" as one that is likely to result in a rule that may:

(1) Have an annual effect on the economy of $100 million or more or adversely affect in a material way the economy, a sector of the economy, productivity, competition, jobs, the environment, public health or safety, or state, local or tribal government or communities;

(2) Create a serious inconsistency or otherwise interfere with an action taken or planned by another agency;

(3) Materially alter the budgetary impact of entitlements, grants, user fees, or loan programs or the rights and obligations of recipients thereof; or

(4) Raise novel legal or policy issues arising out of legal mandates, the President's priorities, or the principles set forth in the E.O.

Pursuant to the terms of Executive Order 12866, OMB has notified EPA that it considers this a "significant regulatory action" within the meaning of the Executive Order. EPA has submitted this action to OMB for review. Changes made in response to OMB suggestions or recommendations will be documented in the public record.

C. Executive Order 12875

Under Executive Order 12875, EPA may not issue a regulation that is not required by statute and that creates a mandate upon a State, local or tribal government, unless the Federal government provides the funds necessary to pay the direct compliance costs incurred by those governments, or EPA consults with those governments. If EPA complies by consulting, Executive Order 12875 requires EPA to provide to the Office of Management and Budget a description of the extent of EPA's prior consultation with representatives of affected State, local and tribal governments, the nature of their concerns, copies of any written communications from the governments, and a statement supporting the need to issue the regulation. In addition, Executive Order 12875 requires EPA to develop an effective process permitting elected officials and other representatives of State, local and tribal governments "to provide meaningful and timely input to the development of regulatory proposals containing significant unfunded mandates."

EPA has concluded that this rule may create a nominal mandate on State, local or tribal governments and that the Federal government will not provide the funds necessary to pay the direct costs incurred by these governments in complying with the mandate. Specifically, some public entities may be covered sources and will have to add the new data elements to their RMP. In developing this rule, EPA consulted with state, local and tribal governments to enable them to provide meaningful and timely input in the development of this rule. Even though this rule revises Part 68 in a way that does not significantly change the burden imposed by the underlying rule, EPA has taken efforts to involve state and local entities in this regulatory effort. Specifically, much of the rule responds to issues raised by the Electronic Submission Workgroup discussed above, which includes State and local government stakeholders. In addition, EPA has recently conducted seminars with tribal governments; however, there were no concerns raised on any issues that are covered in this rule. EPA discussed the need for issuing this regulation in sections II and III in this preamble. Also, EPA provided OMB with copies of the comments to the proposed rule.

D. Executive Order 13045

Executive Order 13045: "Protection of Children from Environmental Health Risks and Safety Risks" (62 FR 19885, April 23, 1997) applies to any rule that: (1) is determined to be "economically significant" as defined under E.O. 12866, and (2) concerns an environmental health or safety risk that EPA has reason to believe may have a disproportionate effect on children. If the regulatory action meets both criteria, the Agency must evaluate the environmental health or safety effects of the planned rule on children, and explain why the planned regulation is preferable to other potentially effective and reasonably feasible alternatives considered by the Agency.

This final rule is not subject to the E.O. 13045 because it is not "economically significant" as defined in E.O. 12866, and because it does not involve decisions based on environmental health or safety risks.

E. Executive Order 13084

Under Executive Order 13084, EPA may not issue a regulation that is not required by statute, that significantly or uniquely affects the communities of Indian tribal governments, and that imposes substantial direct compliance costs on those communities, unless the Federal government provides the funds necessary to pay the direct compliance costs incurred by the tribal governments, or EPA consults with those governments. If EPA complies by consulting, Executive Order 13084 requires EPA to provide to the Office of Management and Budget, in a separately identified section of the preamble to the rule, a description of the extent of EPA's prior consultation with representatives of affected tribal governments, a summary of the nature of their concerns, and a statement supporting the need to issue the regulation. In addition, Executive Order 13084 requires EPA to develop an effective process permitting elected and other representatives of Indian tribal governments "to provide meaningful and timely input in the development of regulatory policies on matters that significantly or uniquely affect their communities."

Today's rule does not significantly or uniquely affect the communities of Indian tribal governments. Two of the amendments made by this rule, the addition of RMP data elements and the conversion of SIC codes to NAICS codes, impose only minimal burden on any sources that may be owned or operated by tribal governments, such as drinking water and waste water treatment systems. The third amendment made by this rule addresses the procedures for submission of confidential business information in the RMP. The sources that are mentioned above handle chemicals that are known to public (e.g., chlorine for use of disinfection, propane used for fuel, etc.). EPA does not, therefore, expect RMP information on these types of processes to include CBI, so any costs related to CBI will not fall on Indian tribal governments. Accordingly, the requirements of section 3(b) of Executive Order 13084 do not apply to this rule.

Notwithstanding the non-applicability of E. O. 13084, EPA has recently conducted seminars with the tribal governments. However, there were no concerns raised on any issues that are covered in this rule.

F. Regulatory Flexibility

EPA has determined that it is not necessary to prepare a regulatory flexibility analysis in connection with this final rule. EPA has also determined that this action will not have a significant economic impact on a substantial number of small entities. Two of the amendments made by this rule, the addition of RMP data elements and the conversion of SIC codes to NAICS codes, impose only minimal burden on small entities. Moreover, those small businesses that claim CBI when submitting the RMP will not face any costs beyond those imposed by the existing CBI regulations. Even considering the costs of CBI substantiation, however, there is no significant economic impact on a substantial number of small entities. EPA estimates that very few small entities (approximately 500) will claim CBI and that these few entities represent a small fraction of the small entities (less than 5 percent) affected by the RMP rule. Finally, EPA estimates that those small businesses filing CBI will experience a cost which is significantly less than one percent of their annual sales. For a more detailed analysis of the small entity impacts of CBI submission, see Document Number, IV-B-02, available in the docket for this rulemaking (see ADDRESSES section).

G. Paperwork Reduction

1. General

The information collection requirements in this rule have been submitted for approval to the Office of Management and Budget (OMB) under the Paperwork Reduction Act, 44 U.S.C. 3501 *et seq.* An Information Collection Request (ICR) document has been prepared by EPA (ICR No. 1656.05) and a copy may be obtained from Sandy Farmer, by mail at Office of Policy, Regulatory Information Division, U.S. Environmental Protection Agency (2137), 401 M St. SW, Washington, DC 20460, by e-mail at farmer.sandy@epamail.epa.gov or by calling (202) 260–2740. A copy may also be downloaded off the Internet at http://www.epa.gov/icr. The information requirements are not effective until OMB approves them.

The submission of the RMP is mandated by section 112(r)(7) of the CAA and demonstrates compliance with Part 68 consistent with section 114(c) of the CAA. The information collected also will be made available to state and local governments and the public to enhance their preparedness, response, and prevention activities. Certain information in the RMP may be claimed as confidential business information under 40 CFR Part 2 and Part 68.

This rule will impose very little burden on affected sources. First, EPA estimates that the new data elements will require only a nominal burden, .25 hours for a typical source, because latitude and longitude method and description will be selected from a list of options, the Title V permit number is available to any source to which Title V applies, and the percentage weight of a toxic substance in a liquid mixture is usually provided by the supplier of the mixture. Second, the NAICS code provision is simply a change from one code to another.[6] Third, as discussed above in the preamble, EPA believes that the CBI provisions of this rule will add no additional burden beyond what sources otherwise would face in complying with the CBI rules in 40 CFR Part 2. The Agency has calculated the burden of substantiations made for purposes of this rule below.

Burden means the total time, effort, or financial resources expended by persons to generate, maintain, retain, or disclose or provide information to or for a Federal agency. This includes the time needed to review instructions; develop, acquire, install, and utilize technology and system for the purposes of collecting, validating, and verifying information, processing and maintaining information, and disclosing and providing information; adjust the existing ways to comply with any previously applicable instructions and requirements; train personnel to be able to respond to a collection of information; search data sources; complete and review the collection of information; and transmit or otherwise disclose the information.

An agency may not conduct or sponsor, and a person is not required to respond to a collection of information unless it displays a currently valid OMB control number. The OMB control numbers for EPA's regulations are listed in 40 CFR Part 9 and 48 CFR Chapter 15.

2. CBI Burden

In the Notice of Proposed Rulemaking for these amendments, EPA proposed to amend existing 40 CFR Part 68 to add two sections which would clarify the procedures for submitting RMPs that contain confidential business information (CBI). As proposed, CBI would be handled in much the same way as it presently is under other EPA programs, except that EPA would require sources claiming CBI to submit documentation substantiating their CBI claims at the time such claims were made and EPA also would not permit CBI claims for certain data elements which clearly are not CBI. Aside from these procedural changes, however, the proposed rule was substantively identical to the existing rules governing the substantiation of CBI claims, presently codified in 40 CFR Part 2.

At the time it proposed these amendments, EPA estimated the public reporting burden for CBI claims to be 15 hours for chemical manufacturers with Program 3 processes, the only kinds of facilities that EPA expects to be able to claim CBI for any RMP data elements. This estimate was premised upon EPA's assessment that it would require 8.5 hours per claim to develop and submit the CBI substantiation and 6.5 hours to complete an unsanitized version of the RMP, for a total of 15 hours. EPA also estimated that approximately 20 percent of the 4000 chemical manufacturers (out of 64,200 stationary sources estimated to be covered by the RMP rule) may file CBI claims (800 sources). The 800 sources represent a conservative projection based on the Agency's experience under EPCRA program. Consequently, the total annual public reporting burden for filing CBI claims was estimated to be approximately 12,000 hours over three years (800 facilities multiplied by an average burden of 15 hours), or an annual burden of 4,000 hours (Information Collection Request No. 1656.04).

a. *Comment received.* EPA received one comment on the ICR developed for the proposed rule, opposing up-front substantiation of any CBI claims. The commenter stated that "[t]his is a major departure from standard EPA procedure, and would impose a substantial and unjustified burden for several years." The commenter further added that up-front substantiation would significantly increase the burden of this rule, and that up-front substantiation unnecessarily increases the volume and potential loss of CBI documents. The commenter also stated that the estimate of 15 hours for chemical manufacturers "seems unreasonably low," and cited the EPA burden estimate of 27.7 to 33.2 hours per claim (with an average of 28.8) under the trade secret provisions of EPCRA.

In the preamble to the proposed rule, EPA estimated that 20 percent of the 4,000 chemical manufacturers will file a CBI claim. The commenter contends that "[t]he EPA analysis * * * excludes facilities in other industries that will need to file CBI claims."

Finally, the commenter stated that claiming multiple data elements as CBI will increase reporting burden.

b. *EPA response. Burden Estimates:* EPA disagrees with these comments. As pointed out above, the requirement to submit up-front substantiation of CBI claims imposes no additional burden. In addition, the total burden of the CBI provisions of this rule are not understated. EPA has re-examined its analysis in light of the commenter's concerns and has determined—contrary to the commenter's claim—that its initial estimate of the total burden associated with preparing and claiming CBI was likely too conservative. As explained below, the Agency's best available information indicates that the process of documenting and submitting a claim of CBI should impose a burden of approximately 9.5 hours per CBI claimant.

First, EPA believes that the requirement to submit, at the time a source claims information as CBI,

[6] EPA intends to provide several outreach mechanisms to assist sources in identifying their new NAICS code. RMP*Submit will provide a "pick list" that will make it easier for sources to find the appropriate code. Also, selected NAICS codes are included in the General Guidance for Risk Management Programs (July 1998) and in the industry-specific guidance documents that EPA is developing. EPA will also utilize the Emergency Planning and Community Right-to-Know Hotline at 800–424–9346 (or 703–412–9810) to assist sources in determining the source's NAICS codes.

substantiation demonstrating that the material truly is CBI imposes no burden on sources beyond that which presently exists under EPA's CBI regulations in Part 2. In order to decide whether they might properly claim CBI for a given piece of information, a source must determine if the criteria stated in section 2.208 of 40 CFR Part 2 are satisfied. Naturally, a source goes through this process before a CBI claim is made. EPA agrees that most programs do not require the information that forms the basis for the substantiation to be submitted at the time of the claim; however, a facility must still determine whether or not a claim can be substantiated. Because existing rules require sources to formulate a legitimate basis for claiming CBI, even if those rules do not require immediate documentation, and because the Agency fully expects requests for RMP information which will necessitate sources' submitting such documentation, EPA believes that up-front submission will not increase the burden of the regulation.

Second, in response to the commenter's claim that the Agency had underestimated the total burden associated with CBI claims, EPA undertook a review of recent information collection requests (ICRs) covering data similar to that required to be submitted in an RMP. Initially, EPA examined the ICR prepared for Part 2 itself (ICR No. 1665.02, OMB Control No. 2020–0003). Under an analysis contained in the Statement of Support for the ICR, the Agency estimated that it takes approximately 9.4 hours to substantiate claims of CBI, prepare documentation, and submit such documentation to EPA. Next, the Agency reviewed a survey conducted by the Agency (under Office of Management and Budget clearance #2070–0034), to present the average burden associated with indicating confidential business information claims for certain data elements under the proposed inventory update rule (IUR) amendment under TSCA section 8. This survey specifically asked affected industry how long it would take to prepare CBI claims for two data elements—chemical identity and production volume range information. Part 68 also requires similar information (e.g., chemical identity and maximum quantity in a process) to be included in a source's RMP and, indeed, EPA anticipates that they will be the data elements most likely to be claimed CBI. The average burden estimates for chemical identity were between 1.82 and 3.13 hours, and the average burden estimates for production volume in ranges were between 0.87 and 2.08 hours. Thus, assuming that the average source claims both chemical identity and the maximum quantity in a process as CBI, a conservative estimate for the reporting burden would be 5.21 hours. Finally, EPA examined the burden estimate upon which it relied at proposal. That estimate predicted that the average CBI claim would take 15 hours, of which 8.5 would be developing and submitting the CBI claim, and 6.5 would be completing an unsanitized version of the RMP. In view of EPA's current plan not to require a source claiming CBI to submit a full, unsanitized RMP, but instead to submit only the particular elements claimed as CBI, the Agency expects the latter burden to decrease to 1 hour, for a total burden of 9.5 hours.

In light of its extensive research of the burden hours involved in preparing and submitting CBI claims, EPA believes that the total burden estimate was not understated in the April proposal. Rather, other ICRs and the ICR proposal, combined with the changes to the method of documenting CBI claims, indicate that a burden estimate between 5.21 and 9.5 hours is appropriate for this final rule. EPA has selected the most conservative of these, 9.5 hours, in its ICR for this final rule.

EPA rejected one ICR's burden estimate as being inapplicable to the present rulemaking. Although the commenter urged the Agency to adopt the estimate associated with trade secret claims under EPCRA (28 hours), EPA believes that the estimates discussed above are more accurate for several reasons. First, the EPCRA figures are based upon a survey with a very small sample size, as compared to the TSCA survey cited previously. Second, most (if not all) of the facilities submitting RMPs are likely to already be reporting under sections 311 and 312 or section 313 of EPCRA, and many of the manufacturers submitting an RMP are subject to TSCA reporting requirements; thus, most sources likely to claim CBI for an RMP data element will have already done some analysis of whether or not such information would reveal legitimately confidential matter.

Other Facilities Can Claim CBI: The Agency does not agree with the commenter's claim that facilities other than chemical manufacturers might be expected to claim CBI for information contained in their RMPs. The other industries affected by the RMP rule (e.g., propane retailers, publicly owned treatment works) will not be disclosing in the RMP information that is likely to cause substantial harm to the business's competitive position. For example, covered public drinking water and wastewater treatment plants generally use common regulated substances in standard processes (i.e., chlorine used for disinfection). Also, covered processes at many sources involve the storage of regulated substances that the sources sell (e.g., propane, ammonia), so the processes are already public knowledge. Other covered processes involve the use of well-known combinations of regulated substances such as refrigerants. Therefore, it is not likely that these businesses would claim information as CBI.

As a point of comparison, EPA notes that of the 869,000 facilities that are estimated to be required to report under sections 311 and 312 of EPCRA, approximately 58 facilities have submitted trade secret claims for under those sections. For this reason, EPA believes the estimate of 800 sources may, in fact, be an overestimate of the number of sources claiming CBI.

Reporting Multiple Data Elements: The Agency disagrees with the commenters assertion that it has underestimated the reporting burden on sources' claiming multiple data elements as CBI. The burden figures stated above are based on the Agency's estimates of the average number of data elements that a typical source will likely claim CBI.

Public reporting of the new RMP data elements is estimated to require an average of .25 hours for all sources (64,200 sources) and substantiating CBI claims is estimated to take approximately 9.5 hours for certain chemical manufacturing sources (800 sources). The aggregate increase in burden over that estimated in the previous Information Collection Request (ICR) for part 68 is estimated to be about 23,650 hours over three years, or an annual burden of 7,883 hours for the three years covered by the ICR.

H. Unfunded Mandates Reform Act

Title II of the Unfunded Mandates Reform Act of 1995 (UMRA), P.L. 104–4, establishes requirements for Federal agencies to assess the effects of their regulatory actions on State, local, and tribal governments and the private sector. Under section 202 of the UMRA, EPA generally must prepare a written statement, including a cost-benefit analysis, for proposed and final rules with "Federal mandates" that may result in expenditures to State, local, and tribal governments, in the aggregate, or to the private sector, of $100 million or more in any one year. Before promulgating an EPA rule for which a written statement is needed, section 205 of the UMRA generally requires EPA to identify and consider a reasonable number of regulatory alternatives and adopt the least costly, most cost-effective or least burdensome alternative that achieves the objectives of the rule. The provisions of section 205 do not apply when they are inconsistent with applicable law. Moreover, section 205 allows EPA to adopt an alternative other than the least costly, most cost-effective or least burdensome alternative if the Administrator publishes with the final rule an explanation why that alternative was not adopted. Before EPA establishes any regulatory requirements that may significantly or uniquely affect small governments, including tribal governments, it must have developed under section 203 of the UMRA a small government agency plan. The plan must provide for notifying potentially affected small governments, enabling officials of affected small governments to have meaningful and timely input in the development of EPA regulatory proposals with significant Federal intergovernmental mandates, and informing, educating, and advising small governments on compliance with the regulatory requirements.

EPA has determined that this rule does not contain a Federal mandate that may result in expenditures of $100 million or more for state, local, and tribal governments, in the aggregate, or the private sector in any one year. The EPA has determined that the total nationwide capital cost for these rule amendments is zero and the annual nationwide cost for these amendments is less than $1 million. Thus, today's rule is not subject to the requirements of sections 202 and 205 of the Unfunded Mandates Act.

EPA has determined that this rule contains no regulatory requirements that might significantly or uniquely affect small governments. Small governments are unlikely to claim information confidential, because sources owned or operated by these entities (e.g., drinking water and waste water treatment systems), handle chemicals that are known to public. The new data elements and the conversion of SIC codes to NAICS codes impose only minimal burden on these entities.

I. National Technology Transfer and Advancement Act

Section 12(d) of the National Technology Transfer and Advancement Act of 1995 ("NTTAA"), Pub L. 104–113, section 12(d)(15 U.S.C. 272 note), directs EPA to use voluntary consensus standards in its regulatory activities unless to do so would be inconsistent with applicable law or otherwise impractical. Voluntary consensus standards are technical standards (e.g., materials specifications, test methods, sampling procedures, business practices) that are developed or adopted by voluntary consensus standards bodies. The NTTAA requires EPA to provide Congress, through OMB, explanations when the Agency decides not to use available and applicable voluntary consensus standards.

This action does not involve technical standards. Therefore, EPA did not consider the use of any voluntary consensus standards.

J. Congressional Review Act

The Congressional Review Act, 5 U.S.C. section 801 et seq., as added by the Small Business Regulatory Enforcement Fairness Act of 1996, generally provides that before a rule may take effect, the agency promulgating the rule must submit a rule report, which includes a copy of the rule, to each House of the Congress and to the Comptroller General of the United States. EPA will submit a report containing this rule and other required information to the U.S. Senate, the U.S. House of Representatives, and the Comptroller General of the United States prior to publication of the rule in the Federal Register. This action is not a "major rule" as defined by 5 U.S.C. section 804(2). This rule will be effective February 5, 1999.

APPENDIX TO PREAMBLE—DATA ELEMENTS THAT MAY NOT BE CLAIMED AS CBI

Rule element	Comment
68.160(b)(1) Stationary source name, street, city, county, state, zip code, latitude, and longitude, method for obtaining latitude and longitude, and description of location that latitude and longitude represent.	This information is filed with EPA and other agencies under other regulations and is made available to the public and, therefore, does not meet the criteria for CBI claims. It is also available in business and other directories.
68.160(b)(2) Stationary source Dun and Bradstreet number.	
68.160(b)(3) Name and Dun and Bradstreet number of the corporate parent company.	
68.160(b)(4) The name, telephone number, and mailing address of the owner/operator.	
68.160(b)(5) The name and title of the person or position with overall responsibility for RMP elements and implementation.	This information provides no information that would affect a source's competitive position.
68.160(b)(6) The name, title, telephone number, and 24-hour telephone number of the emergency contact.	This information is filed with state and local agencies under EPCRA and is made available to the public and, therefore, does not meet the criteria for CBI claims.
68.160(b)(7) Program level and NAICS code of the process.	This information provides no information that would affect a source's competitive position.
68.160(b)(8) The stationary source EPA identifier.	This information provides no information that would affect a source's competitive position.
68.160(b)(10) Whether the stationary source is subject to 29 CFR 1910.119.	This information provides no information that would affect a source's competitive position.
68.160(b)(11) Whether the stationary source is subject to 40 CFR Part 355.	Sources are required to notify the state and local agencies if they are subject to this rule; this information is available to the public and, therefore, does not meet the criteria for CBI claims.
68.160(b)(12) If the stationary source has a CAA Title V operating permit, the permit number.	This information will be known to state and federal air agencies and is available to the public and, therefore, does not meet the criteria for CBI claims.

APPENDIX TO PREAMBLE—DATA ELEMENTS THAT MAY NOT BE CLAIMED AS CBI—Continued

Rule element	Comment
68.160(b)(13) The date of the last safety inspection and the identity of the inspecting entity.	This information provides no information that would affect a source's competitive position.
68.165(b)(4) Basis of the results (give model name if used).	Without the chemical name and quantity, this reveals no business information.
68.165(b)(9) Wind speed and atmospheric stability class (toxics only).	This information provides no information that would affect a source's competitive position.
68.165(b)(10) Topography (toxics only)	Without the chemical name and quantity, this reveals no business information.
68.165(b)(11) Distance to an endpoint	By itself, this information provides no confidential information. Other elements that would reveal chemical identity or quantity may be claimed as CBI.
68.165(b)(12) Public and environmental receptors within the distance.	By itself, this information provides no confidential information. Other elements that would reveal chemical identity or quantity may be claimed as CBI.
68.168 Five-year accident history	Sources are required to report most of these releases and information (chemical released, quantity, impacts) to the federal, state, and local agencies under CERCLA and EPCRA; these data are available to the public and, therefore, do not meet the criteria for CBI claims. Much of this information is also available from the public media.
68.170(b), (d), (e)(1), and (f)–(k) 68.175(b), (d), (e)(1), and (f)–(p) NAICS code, prevention program compliance dates and information.	NAICS codes and the prevention program compliance dates and information provide no information that would affect a source's competitive position.
68.180 Emergency response program	This information provides no information that would affect a source's competitive position.

List of Subjects in 40 CFR Part 68

Environmental protection, Administrative practice and procedure, Air pollution control, Chemicals, Hazardous substances, Intergovernmental relations, Reporting and recordkeeping requirements.

Dated: December 29, 1998.

Carol M. Browner,

Administrator.

For the reasons set out in the preamble, title 40, chapter I, subchapter C, part 68 of the Code of Federal Regulations is amended to read as follows:

PART 68—CHEMICAL ACCIDENT PREVENTION PROVISIONS

1. The authority citation for Part 68 continues to read as follows:

Authority: 42 U.S.C. 7412(r), 7601(a)(1), 7661–7661f.

2. Section 68.3 is amended by removing the definition of SIC and by adding in alphabetical order the definition for NAICS to read as follows:

§ 68.3 Definitions.

* * * * *

NAICS means North American Industry Classification System.

* * * * *

3. Section 68.10 is amended by revising paragraph (d)(1) to read as follows:

§ 68.10 Applicability.

* * * * *

(d) * * *

(1) The process is in NAICS code 32211, 32411, 32511, 325181, 325188, 325192, 325199, 325211, 325311, or 32532; or

* * * * *

4. Section 68.42 is amended by revising paragraph (b)(3), redesignating paragraphs (b)(4) through (b)(10) as paragraphs (b)(5) through (b)(11) and by adding a new paragraph (b)(4) to read as follows:

§ 68.42 Five-year accident history.

* * * * *

(b) * * *

(3) Estimated quantity released in pounds and, for mixtures containing regulated toxic substances, percentage concentration by weight of the released regulated toxic substance in the liquid mixture;

(4) Five- or six-digit NAICS code that most closely corresponds to the process;

* * * * *

5. Section 68.79 is amended by revising paragraph (a) to read as follows:

§ 68.79 Compliance audits.

(a) The owner or operator shall certify that they have evaluated compliance with the provisions of this subpart at least every three years to verify that procedures and practices developed under this subpart are adequate and are being followed.

* * * * *

6. Section 68.150 is amended by adding paragraph (e) to read as follows:

§ 68.150 Submission.

* * * * *

(e) Procedures for asserting that information submitted in the RMP is entitled to protection as confidential business information are set forth in §§ 68.151 and 68.152.

7. Section 68.151 is added to read as follows:

§ 68.151 Assertion of claims of confidential business information.

(a) Except as provided in paragraph (b) of this section, an owner or operator of a stationary source required to report or otherwise provide information under this part may make a claim of confidential business information for any such information that meets the criteria set forth in 40 CFR 2.301.

(b) Notwithstanding the provisions of 40 CFR part 2, an owner or operator of a stationary source subject to this part may not claim as confidential business information the following information:

(1) Registration data required by § 68.160(b)(1) through (b)(6) and (b)(8), (b)(10) through (b)(13) and NAICS code and Program level of the process set forth in § 68.160(b)(7);

(2) Offsite consequence analysis data required by § 68.165(b)(4), (b)(9), (b)(10), (b)(11), and (b)(12).

(3) Accident history data required by § 68.168;

(4) Prevention program data required by § 68.170(b), (d), (e)(1), (f) through (k);

(5) Prevention program data required by § 68.175(b), (d), (e)(1), (f) through (p); and

(6) Emergency response program data required by § 68.180.

(c) Notwithstanding the procedures specified in 40 CFR part 2, an owner or operator asserting a claim of CBI with respect to information contained in its RMP, shall submit to EPA at the time it submits the RMP the following:

(1) The information claimed confidential, provided in a format to be specified by EPA;

(2) A sanitized (redacted) copy of the RMP, with the notation "CBI" substituted for the information claimed confidential, except that a generic category or class name shall be substituted for any chemical name or identity claimed confidential; and

(3) The document or documents substantiating each claim of confidential business information, as described in §68.152.

8. Section 68.152 is added to read as follows:

§68.152 Substantiating claims of confidential business information.

(a) An owner or operator claiming that information is confidential business information must substantiate that claim by providing documentation that demonstrates that the claim meets the substantive criteria set forth in 40 CFR 2.301.

(b) Information that is submitted as part of the substantiation may be claimed confidential by marking it as confidential business information. Information not so marked will be treated as public and may be disclosed without notice to the submitter. If information that is submitted as part of the substantiation is claimed confidential, the owner or operator must provide a sanitized and unsanitized version of the substantiation.

(c) The owner, operator, or senior official with management responsibility of the stationary source shall sign a certification that the signer has personally examined the information submitted and that based on inquiry of the persons who compiled the information, the information is true, accurate, and complete, and that those portions of the substantiation claimed as confidential business information would, if disclosed, reveal trade secrets or other confidential business information.

9. Section 68.160 is amended by revising paragraphs (b)(1), (b)(7), and (b)(12) and adding paragraphs (b)(14) through (b)(18) to read as follows:

§68.160 Registration.

* * * * *

(b) * * *

(1) Stationary source name, street, city, county, state, zip code, latitude and longitude, method for obtaining latitude and longitude, and description of location that latitude and longitude represent;

* * * * *

(7) For each covered process, the name and CAS number of each regulated substance held above the threshold quantity in the process, the maximum quantity of each regulated substance or mixture in the process (in pounds) to two significant digits, the five- or six-digit NAICS code that most closely corresponds to the process, and the Program level of the process;

* * * * *

(12) If the stationary source has a CAA Title V operating permit, the permit number; and

* * * * *

(14) Source or Parent Company E-Mail Address (Optional);

(15) Source Homepage address (Optional)

(16) Phone number at the source for public inquiries (Optional);

(17) Local Emergency Planning Committee (Optional);

(18) OSHA Voluntary Protection Program status (Optional);

10. Section 68.165 is amended by revising paragraph (b) to read as follows:

§68.165 Offsite consequence analysis.

* * * * *

(b) The owner or operator shall submit the following data:

(1) Chemical name;

(2) Percentage weight of the chemical in a liquid mixture (toxics only);

(3) Physical state (toxics only);

(4) Basis of results (give model name if used);

(5) Scenario (explosion, fire, toxic gas release, or liquid spill and evaporation);

(6) Quantity released in pounds;

(7) Release rate;

(8) Release duration;

(9) Wind speed and atmospheric stability class (toxics only);

(10) Topography (toxics only);

(11) Distance to endpoint;

(12) Public and environmental receptors within the distance;

(13) Passive mitigation considered; and

(14) Active mitigation considered (alternative releases only);

11. Section 68.170 is amended by revising paragraph (b) to read as follows:

§68.170 Prevention program/Program 2.

* * * * *

(b) The five- or six-digit NAICS code that most closely corresponds to the process.

* * * * *

12. Section 68.175 is amended by revising paragraph (b) to read as follows:

§68.175 Prevention program/Program 3.

* * * * *

(b) The five- or six-digit NAICS code that most closely corresponds to the process.

* * * * *

13. Section 68.180 is amended by revising paragraph (b) to read as follows:

§68.180 Emergency response program.

* * * * *

(b) The owner or operator shall provide the name and telephone number of the local agency with which emergency response activities and the emergency response plan is coordinated.

* * * * *

[FR Doc. 99–231 Filed 1–5–99; 8:45 am]

BILLING CODE 6560–50–P

www.ingramcontent.com/pod-product-compliance
Lightning Source LLC
Chambersburg PA
CBHW080635180526
45168CB00008B/3183
9781514325148